T0238331

IFIP Advances in Information and Communication Technology

325

IFIP – The International Federation for Information Processing

IFIP was founded in 1960 under the auspices of UNESCO, following the First World Computer Congress held in Paris the previous year. An umbrella organization for societies working in information processing, IFIP's aim is two-fold: to support information processing within its member countries and to encourage technology transfer to developing nations. As its mission statement clearly states,

> *IFIP's mission is to be the leading, truly international, apolitical organization which encourages and assists in the development, exploitation and application of information technology for the bene t of all people.*

IFIP is a non-profitmaking organization, run almost solely by 2500 volunteers. It operates through a number of technical committees, which organize events and publications. IFIP's events range from an international congress to local seminars, but the most important are:

- The IFIP World Computer Congress, held every second year;
- Open conferences;
- Working conferences.

The flagship event is the IFIP World Computer Congress, at which both invited and contributed papers are presented. Contributed papers are rigorously refereed and the rejection rate is high.

As with the Congress, participation in the open conferences is open to all and papers may be invited or submitted. Again, submitted papers are stringently refereed.

The working conferences are structured differently. They are usually run by a working group and attendance is small and by invitation only. Their purpose is to create an atmosphere conducive to innovation and development. Refereeing is less rigorous and papers are subjected to extensive group discussion.

Publications arising from IFIP events vary. The papers presented at the IFIP World Computer Congress and at open conferences are published as conference proceedings, while the results of the working conferences are often published as collections of selected and edited papers.

Any national society whose primary activity is in information may apply to become a full member of IFIP, although full membership is restricted to one society per country. Full members are entitled to vote at the annual General Assembly, National societies preferring a less committed involvement may apply for associate or corresponding membership. Associate members enjoy the same benefits as full members, but without voting rights. Corresponding members are not represented in IFIP bodies. Affiliated membership is open to non-national societies, and individual and honorary membership schemes are also offered.

Arthur Tatnall (Ed.)

History
of Computing

Learning from the Past

IFIP WG 9.7 International Conference, HC 2010
Held as Part of WCC 2010
Brisbane, Australia, September 20-23, 2010
Proceedings

 Springer

Volume Editor

Arthur Tatnall
Victoria University
Victoria Graduate School of Business
City Flinders Campus, PO Box 14428, Melbourne, VIC 8001, Australia
E-mail: arthur.tatnall@vu.edu.au

CR Subject Classification (1998): K.4, K.3, I.2.6, F.1.1

ISSN   1868-4238
ISBN-10   3-642-42314-0 Springer Berlin Heidelberg New York
ISBN-13   978-3-642-42314-7 Springer Berlin Heidelberg New York

springer.com

© IFIP International Federation for Information Processing 2010

Softcover re-print of the Hardcover 1st edition 2010
Typesetting: Camera-ready by author, data conversion by Scientific Publishing Services, Chennai, India
Printed on acid-free paper   06/3180

IFIP World Computer Congress 2010
(WCC 2010)

Message from the Chairs

Every two years, the International Federation for Information Processing (IFIP) hosts a major event which showcases the scientific endeavors of its over one hundred technical committees and working groups. On the occasion of IFIPís 50th anniversary, 2010 saw the 21st IFIP World Computer Congress (WCC 2010) take place in Australia for the third time, at the Brisbane Convention and Exhibition Centre, Brisbane, Queensland, September 20–23, 2010.

The congress was hosted by the Australian Computer Society, ACS. It was run as a federation of co-located conferences offered by the different IFIP technical committees, working groups and special interest groups, under the coordination of the International Program Committee.

The event was larger than ever before, consisting of 17 parallel conferences, focusing on topics ranging from artificial intelligence to entertainment computing, human choice and computers, security, networks of the future and theoretical computer science. The conference History of Computing was a valuable contribution to IFIPs 50th anniversary, as it specifically addressed IT developments during those years. The conference e-Health was organized jointly with the International Medical Informatics Association (IMIA), which evolved from IFIP Technical Committee TC-4 "Medical Informatics".

Some of these were established conferences that run at regular intervals, e.g., annually, and some represented new, groundbreaking areas of computing. Each conference had a call for papers, an International Program Committee of experts and a thorough peer reviewing process of full papers. The congress received 642 papers for the 17 conferences, and selected 319 from those, representing an acceptance rate of 49.69% (averaged over all conferences). To support interoperation between events, conferences were grouped into 8 areas: Deliver IT, Govern IT, Learn IT, Play IT, Sustain IT, Treat IT, Trust IT, and Value IT.

This volume is one of 13 volumes associated with the 17 scientific conferences. Each volume covers a specific topic and separately or together they form a valuable record of the state of computing research in the world in 2010. Each volume was prepared for publication in the Springer IFIP Advances in Information and Communication Technology series by the conference's volume editors. The overall Publications Chair for all volumes published for this congress is Mike Hinchey.

For full details of the World Computer Congress, please refer to the webpage at http://www.ifip.org.

June 2010   Augusto Casaca, Portugal, Chair, International Program Committee
Phillip Nyssen, Australia, Co-chair, International Program Committee
Nick Tate, Australia, Chair, Organizing Committee
Mike Hinchey, Ireland, Publications Chair
Klaus Brunnstein, Germany, General Congress Chair

Preface

History of Computing: Learning from the Past

Why is the history of computing important? Given that the computer, as we now know it, came into existence less than 70 years ago it might seem a little odd to some people that we are concerned with its history. Isn't history about 'old things'?

Computing, of course, goes back much further than 70 years with many earlier devices rightly being known as computers, and their history is, of course, important. It is only the history of *electronic digital computers* that is relatively recent.

History is often justified by use of a quote from George Santayana who famously said that: 'Those who cannot remember the past are condemned to repeat it'. It is arguable whether there are particular mistakes in the history of computing that we should avoid in the future, but there is some circularity in this question, as the only way we will know the answer to this is to study our history.

This book contains papers on a wide range of topics relating to the history of computing, written both by historians and also by those who were involved in creating this history. The papers are the result of an international conference on the History of Computing that was held as a part of the IFIP World Computer Congress in Brisbane in September 2010.

All the papers in this book were subjected to a rigorous process of peer review by three reviewers and only accepted for publication after appropriate changes had been made to address reviewers' comments.

Arthur Tatnall

Organization

International Program Committee

Arthur Tatnall (Chair)	Victoria University, Australia
David Anderson	University of Portsmouth, UK
Corrado Bonfanti	Italian Computer Society, Italy
Klaus Brunnstein	University of Hamburg, Germany
Paul Ceruzzi	Smithsonian Institution, USA
Bill Davey	RMIT University, Australia
Graham Farr	Monash University, Australia
Ashley Goldsworthy	Australia
Lars Heide	Copenhagen Business School, Denmark
John Impagliazzo	Qatar University, Qatar
Roger Johnson	University of London, UK
Peter Juliff	Deakin University, Australia
Bill Olle	UK
Kevin Parker	Idaho State University, USA
Judy Sheard	Monash University, Australia
Jeffery Stein	IT History Society, USA
Barrie Thompson	University of Sunderland, UK

Prologue: An Illustrated History of Computing in Australia

Max Burnet, from BACK Pty Ltd, Sydney, Australia, presented a keynote address to the conference. This keynote presentation began with an overview of computation in Australia from about 100 years ago and went on to outline the development of Australia's first electronic computers. (The acronym BACK Pty Ltd stands for Burnet Antique Computer Knowhow.)

1 An Illustrated History

This presentation started with an overview of computation in Australia from about 100 years ago, a period of wonderful brass instruments, ready reckoners and mechanical monsters called totalisators. The presentation outlined the development of Australia's first electronic computers after World War II including the famous CSIRAC, Deuce, Silliac, Snocom, WREDAC and CIRRUS, and covered the levels of Australian content as the industry developed.

Max showed that in the mainframe era, Australia was a fertile ground for sales by UK, Japan and USA manufacturers and outlined the mini-computer era, then the explosion in creativity of the Micro era. He featured Australian products that were unique on the world stage – Dulmont, Microbee etc., and examined how government policies such as Local Content Demands and Offsets Polices have influenced the industry. Max used pictures mainly from his own collection of artifacts as featured on ABC's "The Collectors" show last May (2010). The history finished in about 1985, when Intel, Microsoft and Moore's law made the history of computing a lot less interesting.

Max finally outlined how the Australian papers submitted to the HC conference stream fit into this historic sequence and also mentioned the artifacts present in the historical display at the conference.

2 BACK

Burnet Antique Computer Knowhow Pty Ltd has an extensive collection of early computers and data processing artifacts dating back to 1910. It comprises over 60 six-foot cabinets. It is considered to be the best collection of Digital Equipment Corporation material in the world. There are also over 6,000 items of literature, handbooks, engineering diagrams and software, all catalogued and accessible.

BACK has no grand plan to have these items gathering dust in a museum, but rather to maintain and use them as working tools for a variety of projects. BACK is convinced that in 50 or 100 years, such items will be of immense value and interest.

All these items are currently distributed in a number of warehouses around the Sydney area. Inspections can be arranged. Some are on display in public places such as company foyers etc.

Table of Contents

Connections in the History of Australian Computing

John Deane

Australian Computer Museum Society,
PO Box S-5, Homebush South, NSW 2140, Australia
jdeane@ihug.com.au

Abstract. This paper gives an overview of early Australian computing mile-
stones up to about 1970 and demonstrates a mesh of influences. Wartime radar,
initially from Britain, provided basic experience for many computing engineers.
UK academic Douglas Hartree seems to have known all the early developers
and he played a significant part in the first Australian computing conference.
John von Neumann's two pioneering designs directly influenced four of the first
Australian machines, and published US designs were taken up enthusiastically.
Influences passed from Australia to the world too. Charles Hamblin's *Reverse
Polish Notation* influenced English Electric's *KDF9*, and succeeding stack ar-
chitecture computers. Chris Wallace contributed to English Electric, and
Murray Allen worked at Control Data. Of course the Australians influenced
each other: Myers, Pearcey, Ovenstone, Bennett and Allen organized confer-
ences, interacted on projects, and created the Australian Computer Societies.
Even horse racing played a role.

Keywords: Australia, computing, Myers, Pearcey, Ovenstone, Bennett, Allen,
Wong, Hamblin, Hartree, Wilkes, CSIRO, CSIRAC, SILLIAC, UTECOM,
WREDAC, SNOCOM, CIRRUS, ATROPOS, ARCTURUS.

1 Introduction

The history of computing in Australia can be seen as a nearly continuous series of
personal connections, both within the country and internationally. The following sections highlight some of the influences on the early Australian projects.

2 George Julius' Automatic Totalisator

There have been calculating aids in Australia for as long as there have been people
here, but one of the first mechanical aids associated with Australia was developed by
a mechanical engineer, George Julius. George was born in England[1], brought up in
New Zealand, and it might be fair to say he was obsessed with mechanical gadgets.
From 1896 he was working as a railway engineer in Western Australia and

---

[1] Born in Norwich, England 29 April 1873, died 28 June 1947.

A. Tatnall (Ed.): HC 2010, IFIP AICT 325, pp. 1–12, 2010.

*"A friend in the west conceived the idea of getting me to make a machine to regis-
ter votes, and so to expedite elections by giving the result without any human
intervention."* [1]

George liked the challenge and by 1906 he had a "Voting machine"[2] which he offered
to the Australian Federal Government. They could see difficulties, and considerable
expense, and declined. But George's advisers would not be thwarted:

*"A friend who knew of a 'jam tin tote' - a machine which kept a sort of record of
tickets sold at each window ... I found the problem of great interest as the perfect
tote must have a mechanism capable of adding the records from a number of
operators all of whom might issue a ticket on the same horse at the same instant."*
[2]

George was head-hunted from the railways by a Sydney engineering firm, and also set
himself up as a consulting engineer in 1907. Even with two jobs and a young family
he built a prototype of his Automatic Totalisator in his home workshop [3].[3]

The first *Tote* was installed in Auckland NZ in 1913. Though racing officials predicted that the "giant tangle of piano wires, pulleys and cast iron boxes" wouldn't
work, it was a great success. George patented it in 1914,[4] installed the second in
Western Australia in 1916, set up Automatic Totalisators Ltd. in 1917 and never
looked back. By 1970 Australian Totalisators were in service in 29 countries [4].

The London Science Museum has described the Julius Tote as the "earliest online,
real-time, data-processing and computation system" that the curators can identify [5].

George took part in an increasing number of engineering and scientific bodies, and
when the *Council for Scientific and Industrial Research* (later CSIRO) was formed in
1926 he became its first Chairman [6]. The Council was formed to direct research into
agriculture but by the mid 1930s Sir George became convinced that research should
be extended to industry, and he was instrumental in creating the Aeronautical Research Laboratories, Electrotechnology Section, and the Radiophysics Laboratory in
1939.

3 David Myers' CSIR Differential Analyser

About 1925 George Julius gave a lecture on calculating machines where he demonstrated his Totalisator model [7]. In the audience 14 year old David Myers[5] was very
impressed, and the session generated a lifelong interest in computing machinery.
David graduated from Sydney University then headed for the UK in 1933. Before
starting his PhD at Oxford he briefly worked for a company constructing a mechanical "computer" (a Bush Differential Analyser) for Prof Douglas Hartree at Manchester. David used this D-A later, and maintained contact with Prof Hartree for many
years.

David returned to Sydney in 1936, joined the CSIR *National Standards Laboratory*,
and subsequently headed its new Electrotechnology Section. Instead of standards, this

---

[2] UK patent 1906/28,335.
[3] This model is now in the Sydney Powerhouse Museum collection.
[4] Australian Patent 15133/14, 21 December 1914.
[5] Born Sydney 5 June 1911, died Sydney 11 November 1999 [8].

group was soon pressed into war service developing gun aiming computer devices, and they continued this work with Radiophysics' radar group.

After WW2 David recommended increased work on computing devices and the *Mathematical Instrument Section* was created within his Division. Their first job was the design and construction of an advanced electro-mechanical Differential Analyser for the CSIR. This used technology developed for gun aiming, and was used heavily through the 1950s [9].

4 Trevor Pearcey and CSIRAC

In 1940, a young mathematician, Trevor Pearcey,[6] joined the UK war effort instead of starting a PhD. He worked with the British *Radar Research Establishment* on the way radio signals travelled in different atmospheric conditions. Initially this involved months of work on manual calculators, but after contact with Prof Douglas Hartree the work continued on the Differential Analysers at Manchester then at Cambridge [10, 11].

Late in 1944 the radar research was winding down and Trevor answered a newspaper advertisement for a mathematical physicist with the CSIR Division of Radiophysics. He was successful and, a year later with the end of the war, he sailed for Australia. Trevor accommodated a growing interest in calculating machines by arranging to visit Boston and Howard Aiken's enormous *Automatic Sequence Controlled Calculator* at Harvard University, as well as the latest Bush Differential Analyser nearby at the Massachusetts Institute of Technology.

Trevor established a Mathematical Section in Sydney, then managed to convince management to redirect his radio studies into the development of an electronic computing machine. He argued that this would clearly be needed by Radiophysics' other projects.

The logical design work started in 1946 with radar engineer Maston Beard producing circuit details. While the influences still have to be teased out, Trevor could hardly not have known of UK and US work. In mid 1945 Prof Hartree had prepared a large report into calculating machines [12] which included extensive extracts from John von Neumann's EDVAC report [13], and outlines of the critical memory design. Anyhow, Trevor clearly approached his design from first principles, the essential design of the *CSIR Mark 1* was completed in early 1948 and construction started. A review paper written then started with Charles Babbage and ended with a prediction:

> "in the non-mathematical field there is scope for the use of the [computing] techniques in such things as filing systems. It is not inconceivable that an automatic encyclopaedic service operated through the national teleprinter or telephone service will one day exist." [14]

Radiophysics staff developed major sub-systems and carried out the construction. The Mark 1 ran its first program late in 1949 though the arithmetic unit was not complete [15].

While Trevor and his team worked on their computer, David Myers organised Australia's first *Conference On Automatic Computing Machines* for 1951. Four presentation were given by Prof Hartree, three by David Myers and four by Trevor Pearcey.

---

[6] Born London 5 March 1919, died Melbourne 27 January 1998.

Also there was a substantial equipment display including the *CSIRO Differential Analyser*, manual calculators, considerable punched card equipment - and - the *CSIRO Mark 1 Electronic Computer* [16].

The computer was heavily used, a high-level language was developed alongside the world's first computer music program [17]. Trevor wanted to start on *Mark 2* and CSIRO approached Australian industry for commercial support. This was not forthcoming and CSIRO did not want to fund computer development. After considerable agonising the *Mark 1* was transferred to the University of Melbourne and renamed *CSIRAC* in 1956. There it provided a successful service until 1964 [18].

CSIRAC was preserved by Museum Victoria, it received Heritage Status in 2009, and is believed to be the only intact first generation computer [19].

5   John Ovenstone and WREDAC

Immediately after WW2, and in the shadow of German terror weapons, Britain and Australia agreed to develop rocket technology at a munitions factory north of Adelaide. A corridor was allocated from there stretching 1,800 km north-west to the ocean near Broome in WA, and this test range was christened *Woomera* [20].

As the *Long Range Weapons Establishment* (LRWE) installed tracking cameras, radar, recording equipment and started rocket test firings from 1949, it quickly became obvious that their room full of girls with desk calculators took too long to process the flight data. They knew of Trevor Pearcey's work and sent a group to the 1951 computer conference. Shortly after that they started building a copy of the *CSIRO Mark 1* as the *LRWE Electronic Digital Automatic Computer*, or *LEDAC* [21].

Management had a change of heart, cancelled *LEDAC*, and directed LRWE to purchase a *Ferranti Mark 1*. However, the Ferranti machine wasn't ready, availability kept slipping and its price kept rising.

Also in 1951 LRWE hired a brilliant maths graduate, John Allen-Ovenstone, and sent him to Cambridge to do a doctorate under Douglas Hartree. John had used the CSIRO computer and reached Cambridge shortly after their first computer, *EDSAC*[7], came into operation. When he returned late in 1953 nothing had changed, and he wrote a detailed specification for the computer that LRWE needed [22].

John visited the UK and found Ferranti's *PEGASUS* too complicated, and English Electric's *DEUCE* wasn't ready, but Elliotts was willing to build a special version of their 400 series. Their internal "odd jobs" code, "403", became LRWE's computer. Initially called "Cobber", then the *Elliott 403*, and following LRWE's name change to the *Weapons Research Establishment* (WRE), it was *WREDAC*. The CPU was shipped in mid 1955 and the output processor some months later. By late 1956 *WREDAC* was working well [23].

John Ovenstone managed *WREDAC* but he saw a much bigger picture. He organised a week long computer conference in June 1957 at WRE. There were 25 papers on programming, 23 on engineering, and from John's vision: 15 on business applications. There were demonstrations of *WREDAC* and their analogue computer, Elliotts'

---

[7] *Electronic Delay Storage Automatic Calculator* largely based on von Neumann's work [13].

$AGWAC^8$. Presenters came from London, Cambridge, Leeds and Manchester Universities, the UK National Physical Laboratory, Ferranti, EMI, Elliotts and English Electric. Also Michigan University, the US Cape Canaveral testing ground, three Australian Universities, CSIRO and, of course, WRE. A few of the visitors were Andrew Booth, Stanley Gill, Tom Kilburn, and Maurice Wilkes. Locals including Trevor Pearcey, Murray Allen, John Bennett, Brian Swire and John Ovenstone renewed acquaintances and a great deal of information was transferred [24].

"The conference was a great success, both technically and socially" [25]

WRE hired an *IBM 7090* from 1960, and in late 1962 *WREDAC* was scrapped [26].

6 Brian Swire, John Bennett and SILLIAC

In 1952 the University of Sydney's Physics Department got a new Head, Dr Harry Messel, and a new budget. Messel appointed Dr John Blatt from the University of Illinois. Dr Blatt had programmed Illinois' *ILLIAC*, one of a series of copies of John von Neumann's *IAS Computer*, and easily made the case that Physics needed a computer. Australia's only computer was on the Sydney Uni site, but it was being kept very busy with CSIRO work. Prof Messel campaigned for funding and in 1954 Adolph Basser donated his horse's Melbourne Cup winnings. Dr Blatt arranged to get circuit details and construction samples from Illinois while Prof Messel arranged for staff to build their computer and to program it [27].

Construction of "Sydney's ILLIAC", or *SILLIAC*, was directed by Brian Swire. Brian had worked as a radar engineer at Radiophysics then moved to the CSIRO Aeronautical Research Laboratory, and he attended the 1951 conference. He took the von Neumann/*ILLIAC* design and reworked it for maximum reliability [28].

When they advertised for a software expert who could also teach programming, the best applicant was an Australian working for Ferranti UK. John Bennett had experience with CSIR's radar team and David Myers before heading to the UK. He was Maurice Wilkes first PhD student and he had helped build Cambridge University's first computer *EDSAC*, which was largely based on John von Neumann's *EDVAC* design. John Bennett's work for Ferranti involved reworking the instruction set of their first computer, logic design (including *NIMROD*, the first games console), software development and customer relations [29].

SILLIAC's circuitry was constructed by Sydney electronics firm *Standard Telephones and Cables* then assembled and tested by Brian's small team from mid 1955. John started programming courses, which included a helping of numerical analysis concepts, and assembled the operating software based on Illinois' experience. The first successful run in July 1956 gave Sydney a brief lead in quantum theory[9] and heralded a decade of intense work for the University, CSIRO and business [30].

Late in its life *SILLIAC* was interconnected with other University computers as an input/output server in what we would now call a local area network. Hardware and software were developed by Chris Wallace. Chris later spent some time with English Electric and contributed to their *KDF9* team [31].

---

[8] The *Australian Guided Weapons Analogue Computer* was used to model missile behavior.
[9] Specifically, the mathematical description of helium superfluidity.

When *SILLIAC* was finally turned off in 1968, parts were given to a variety of people, including 14 schoolchildren who wrote in asking for mementos [32].

John Bennett continued lecturing, supporting computing as a profession and, with Trevor Pearcey, founded the Australian Computer Society.

7 UTECOM

At the same time as SILLIAC was being built, Sydney's other university, the New South Wales University of Technology, received a large grant from the state government to study nuclear power. The new head of Electrical Engineering, Rex Vowels, proposed purchasing a computer to support multiple disciplines [33]. Government policy meant the purchase had to be British, and in 1954 that meant a Ferranti *Mark 1*, Lyons *LEO 1* or English Electric *DEUCE*. They felt that the *DEUCE*, derived from Alan Turing's *ACE* design, was the most advanced and ordered the *University of Technology Electronic Computer*, ie *UTECOM*. This was shipped from the UK in mid 1956 and it was used from September [34].

Work on *UTECOM* paralleled *SILLIAC* with intense student, research and commercial activity. An early user was Professor of Philosophy (and ex radar engineer) Charles Hamblin. At the 1957 computer conference he presented a somewhat abstruse maths method "Reverse Polish Notation" and showed how this simplified programming and even hardware design. English Electric engineers at the conference understood the significance and their next major machine, the *KDF9*, used its pushdown/pop-up stack memory extensively [35].

UTECOM went through two rounds of upgrades and its owner changed its name to the University of NSW before it was replaced by an *IBM 360* in 1966, and mostly scrapped.

8 Murray Allen's ADA

In 1949 the Australian government started the Snowy Mountains Hydro-electricity project. This required an unprecedented level of engineering design and, specifically, mathematical modeling of the overall system. A first attempt at this by manual calculation had taken "many man-years", and they needed to do a whole series [36].

The Snowy folk approached David Myers at the CSIRO *Section for Mathematical Instruments* (SMI) and found that they already had a project underway that seemed a good match. The SMI had a brilliant young engineering graduate, Murray Allen, who was developing a new computer as his PhD project. This used a new electronic device, the transistor, which promised speed, reliability, heat, and size advantages over vacuum-tube technology. Murray's project was to rework the US *Bendix D-12 Differential Analyser* design using transistors.[10] This was electronic, digital, programmable and automatic - it was *ADA*, the *Automatic Differential Analyser* [37].

Construction started in 1956 when reliable transistors became commercially available.[11] Adolph Basser contributed funding and the Snowy Authority wanted *ADA-2*

---

[10] The D-12 was based on Northrop's 1950 *Magnetic Drum Digital Differential Analyzer* (MADDIDA).

[11] Philco surface-barrier germanium transistors.

for their exclusive use. *ADA* was ceremonially opened in March 1958 - but its life was short. Its memory was a CSIRO built magnetic drum:

"One afternoon tea [early in 1961 [38]] there was a mighty crash and the drum was essentially destroyed - a piece of lint had lodged under a head and dug a great channel. ADA was done for." [39]

9 David Wong's SNOCOM

During the development of *ADA* one of Sydney Uni's graduate students, David Wong, was given the task of determining what the *Snowy Mountains Hydro-electric Authority* (SMHA) really wanted to do, and what *ADA-2* should be capable of David concluded that while *ADA* had 60 integrators, the full problem would require 400. Also, the SMHA had many non-differential computing jobs, and he showed that while these could be expressed in differential form, it was complicated. David went one step further and programmed a representative differential problem on Sydney University's digital computer *SILLIAC*. His conclusion that a general-purpose computer would do, plus its specification, earned his Master's degree, and a PhD project - to build it [40].

There were some constraints on the project: little money, little time, and little help. Then, in early 1957, the design of a small commercial US computer, the *LGP-30*, was published by its designer, Stanley Frankel [41]. The *Librascope General Purpose computer* used valve logic and a drum memory and was surprisingly similar, in a general way, to *ADA*. David, with Murray Allen, set about expanding the *LGP-30's* description to a design they could build with the modules developed for *ADA* [42].

A simulator for the Snowy Computer - *SNOCOM* - was written on *SILLIAC* [43] and software development started. Much was done through a 500 km teleprinter link from SMHA headquarters in Cooma!

SNOCOM was delivered to SMHA at Cooma in August 1960. By 1962, 50 programmers kept *SNOCOM* busy for two shifts a day, and it was augmented by an Elliott computer. In 1967 *SNOCOM* was retired to student work at Sydney Uni, then presented to the Powerhouse Museum [44].

10 Murray Allen, Trevor Pearcey and CIRRUS

Following the transfer of *CSIRAC* to the University of Melbourne a very disappointed Trevor Pearcey returned to the UK *Radar Research Establishment* (RRE) late in 1957. Their fast, and largely secret, vacuum-tube computer *TREAC* had been operating since 1953 and Trevor worked on compilers and a subroutine library stored in read-only ferrite-rod memory. He also had contact with Maurice Wilkes and the brand new *EDSAC 2* at Cambridge. This also had read-only memory, here used to control instruction execution - it was the first microprogrammed computer [45, 46].

In 1959 Trevor returned to Australia and joined the CSIRAC Laboratory in Melbourne. At the same time Murray Allen left Sydney University and joined the University of Adelaide to establish their computer department. Murray got his staff and students thinking about a big computer project and he talked to Trevor and the Weapons Research computing team. Their initial goal was to produce a cheap, open-ended architecture with minimal hardware based around ferrite core memory and related read-only store. They dubbed the "blue sky", unfunded project *CIRRUS*, and proceeded to produce a detailed design. The instruction microprograms were simulated by Trevor on *CSIRAC* and software design started [47].

Adelaide University saw a very promising future for the design, supported the project and stimulated funding from the Postmaster General, and Weapons Research Establishment.

The basic hardware elements that Murray used in *ADA* and *SNOCOM* were extended for *CIRRUS* and constructed by a PMG contractor. As the hardware developed it became clear that it would be so fast that a form of multiprogramming would be needed to keep the processor busy. The team knew of a few commercial examples (*Honeywell 800* and *IBM 7030 Stretch*) but their requirements were different. John Penny wrote a simulator on WRE's *IBM 7090* and developed the operating software there [48].

CIRRUS came into operation in late 1963 with 4 user workstations. It was fast, inexpensive, and well ahead of commercial contemporaries. It was heavily used despite the installation of a CSIRO *CDC 3200* and a *CDC 6400*. As well as teaching and research programming, special purpose on-line control and signal-processing workstations were built. A prolonged breakdown in 1969 triggered the purchase of a *Data General Nova*, and *CIRRUS* was taken out of service late in 1971 [49].

CIRRUS is preserved in Adelaide Uni's Electrical Engineering Department.

Australian industry did not take up this cheap and flexible design and Adelaide Uni did not pursue computer development. Murray Allen spent a year with Control Data in the USA contributing to the *CDC 3000* series computers, then he moved to the University of NSW. Trevor Pearcey and John Penny joined the new CSIRO Division of Computing Research.

11 ATROPOS

The Weapons Research folk needed to know where the rockets they were testing were likely to come down. They had an "impact predictor" system to do this[12], but with the 1958 *Blue Streak* project the impact point would shift by 60 km every second, and they decided a digital system was needed. It had to run the 5,000 instruction procedure five times a second and there wasn't a commercial system WRE could afford which would do that [50]. A small team of WRE engineers (Ian Hinckfuss, Ron Keith and Ian Macaulay) visited the UK in 1957 and included RRE where Trevor Pearcey was working on *TREAC* [51]. This machine was fast, with parallel operation and very good arithmetic [52]. They felt they could do even better with transistors instead of valves, and core memory instead of William's Tube CRTs.

They had a design by 1960 and WRE workshops constructed their computer by the end of 1962. This was large for a second generation machine: 6m wide, 2m high and ½m deep. Transporting it 150 km from the labs near Adelaide to the Woomera Range was done on an air cushioned truck, at 15 km/h.

Its major input was from two radar units 200 km away, so they also had to invent a reliable data transmission system [53]. Its job was to convert radar data to position

---

[12] Rocket tracking radar was linked to plotting tables watched by the Safety Officer.

and speed, carry out processing quality checks and compare the predicted impact point to the rocket range boundaries. The final stage could be to instruct an errant missile to self-destruct, so it was named after the Greek god of death, *ATROPOS*.

They also wrote a simulator and tested their software on WRE's *IBM 7090*. Then they generated test radar data there too and *ATROPOS* was placed in operation at the end of 1963 [54]. In 1964 it was

> *"one of the few, and certainly the largest real time data processing system operating in Australia. Yet with the exception of the radars and plotting tables, all the equipment involved was designed by WRE and built either in the WRE or by local industry"* [55]

The system was successful and reliable, and was operated until 1974 when it was replaced by a commercial computer [56].

12 ARCTURUS

Following David Wong's completion of *SNOCOM* he built a large digital trainer, *NIMBUS*, and from the early 1960s started thinking about a very economical, general purpose computer for the Electrical Engineering Department at Sydney Uni. There was no budget at all but David started small related projects: a printer controller, and a remarkably fast paper tape reader. He also started salvaging electronic components and developing the design with *SNOCOM* style hardware. By mid 1964 he had a complete design, and a small amount of funding to build a computer for educational use [57].

David bought core memory, a paper tape punch, and the necessary components to construct his machine in the University. It was completed in 1966 at a cost around £5,000. There was also a competition to name it, and *ARCTURUS*[13] won [58].

While there were other computers available *ARCTURUS* was fast and convenient, and its hardware could be modified for special projects. For example, when IBM donated a large disc drive (a *RAMAC 305*) in 1971, it was easiest to build an interface for *ARCTURUS* (which was done by Kevin Rosolen) [59].

By 1975 small and cheap commercial computers designed for digital control were available and *ARCTURUS* was replaced by a Digital Equipment *PDP-11/45* [60].

13 Summing Up

History tends to make champions of inventors and early Australian computing has a notable number of inventions. This brief overview of the obvious early milestones shows quite a mesh of influences. British radar provided basic electronic techniques, the critical delay-line memory device and an excellent reason to automate calculations. UK academic Douglas Hartree knew David Myers, Trevor Pearcey and John Ovenstone. John von Neumann's two fundamental designs, *EDVAC* and the *IAS Computer*, directly influenced *CSIRAC* and *SILLIAC*, and had surprising influences on *WREDAC* and *CIRRUS* via Cambridge's *EDSAC*. Published US designs were

---

[13] In Greek mythology Zeus created Arcturus to guard the bear.

interpreted for *ADA* and *SNOCOM* and further British radar work on *TREAC* influenced *CIRRUS* and *ATROPOS*.

Influences passed out of Australia too. Charles Hamblin's applied mathematical philosophy influenced English Electric's *KDF9*, and arguably all the succeeding stack architecture computers. Chris Wallace worked at English Electric, Trevor Pearcey contributed to *TREAC* and Murray Allen sojourned at Control Data. Exactly what did they contribute?

Of course the Australians influenced each other. David Myers, Trevor Pearcey, John Ovenstone, John Bennett and Murray Allen organized conferences, bounced ideas off each other, educated a generation of engineers and programmers, and did much to create the state and national computer societies.

Finally, this is part of the story up to about 1970. There is of course more, eg *CSIROnet*, Owen Hill's *Microbee*, Alan Bromley's[14] part in the <u>first</u> build of Charles Babbage's *Difference* Engine and the CSIRO Wireless LAN saga!

Appendix - Brief Specifications

System	Used	Brief Specification
Automatic Totalisator	1913 on	Electro-mechanical decimal adders with smart ticket machine scanners. Not programmable.
CSIR Diff. Analyser	1946 to c.1960	Electro-magnetic analogue integrators. Programmed by cable interconnections.
CSIR Mark 1 (CSIRAC)	1949 to 1964	Serial digital computer, vacuum tube logic, mercury delay-line memory, magnetic drum, paper tape i/o.
WREDAC (Elliott 403)	1955 to 1962	Serial digital computer, vacuum tube logic, Ni delay-line memory, mag. disc, mag. tape, paper tape, plotter.
SILLIAC	1956 to 1968	Parallel digital computer, vacuum tube logic, Williams' tube CRT memory, magnetic tape, paper tape i/o.
UTECOM (EE DEUCE)	1956 to 1966	Serial digital computer, vacuum tube logic, mercury delay-line memory, magnetic drum, punched card i/o.
ADA	1958 to 1961	Serial digital differential analyzer, transistor logic, magnetic drum memory, paper tape, plotter.
SNOCOM	1960 to 1967	Serial digital computer, transistor logic, magnetic drum memory, paper tape i/o.
CIRRUS	1963 to 1971	Parallel digital computer, micro-programmed transistor logic, ferrite core memory, multi-programmed operating software in ROM, paper tape i/o.
ATROPOS	1963 to 1974	Parallel digital computer, transistor logic, very fast ×, ÷, √, ferrite core memory, paper tape & parallel i/o.
ARCTURUS	1966 to 1975	Parallel digital computer, transistor logic, ferrite core memory, paper tape i/o.

---

[14] Born 1947, died 16 August 2002.

References

1. Anderson, M., Cochrane, P.: Julius Poole & Gibson: the first eighty years from Tote to CAD. Julius Poole & Gibson, Sydney (1989)
2. ibid
3. ibid
4. Conlon, B.: Automatic Totalisators Limited - later ATL, http://members.ozemail.com.au/~bconlon/atl.htm (accessed 29/11/2009)
5. Swade, D.: Science Goes to the Dogs: a Sure Bet for Understanding Computers. In: New Scientist, October 29 (1987)
6. CSIRO Chairmen - April 1926 to 31 December 1945, http://www.csiro.au/resources/CSIROChairmen.html (accessed 11/11/2009)
7. Anderson, M., Cochrane, P.: ibid (1989)
8. Service for founding Vice-Chancellor. In: Agora La Trobe Uni. Alumni Assoc. Newsletter Autumn (2000), http://www.latrobe.edu.au/alumni/agora/archives/agora_autumn 2000.pdf
9. Bennett, J.M., et al. (eds.): Computing in Australia: the development of a profession, pp. 9–14. Hale & Iremonger, Sydney (1994)
10. Personal details generously provided by Frances Boyd née Pearcey (July 2004)
11. Pearcey, T.: A History of Australian Computing. Chisholm Institute of Technology, Melbourne (1988)
12. Hartree, D.: U.S. Developments in Calculating Machines, National Archives of the History of Computing NAHC/HAR/C1 (July 1945)
13. Von Neumann, J.: First Draft of a Report on the EDVAC. In: Randall, B. (ed.) 1973: The Origins of Digital Computers. Springer, New York (1945)
14. Pearcey, T.: Modern Trends in Machine Computation. Aust. J. Science X/4 Supp. (1948)
15. Pearcey, T.: ibid (1988)
16. Exhibition of Equipment and Demonstrations. In: Proceedings of Conference on Automatic Computing Machines held in the Department of Electrical Engineering, University of Sydney August 1951. CSIRO, Melbourne (April 1952)
17. Doornbusch, P.: The Music of the CSIRAC: Australia's First Computer Music. Common Ground (2005)
18. Willis, J.B., Deane, J.F.: Trevor Pearcey and the First Australian Computer: A Lost Opportunity? In: Historical Records of Australian Science. CSIRO (2006)
19. Victorian Heritage Database - CSIRAC, http://vhd.heritage.vic.gov.au/places/heritage/114928
20. Morton, P.: Fire across the desert: Woomera and the Anglo-Australian Joint Project 1946-1980. AGPS Press, Canberra (1989)
21. Personal recollections from Peter Goddard (2007)
22. Allen-Ovenstone, J.: Notes on data processing at LRWE. NAA A427/2/2 folio 7A (1953)
23. Morton, P.: ibid (1989)
24. WRE 1957: Data Processing and Automatic Computing Machines - Proceedings of Conference held at Weapons Research Establishment, Salisbury S.A. Commonwealth of Australia Dept. of Supply, Adelaide (June 3-8 1957)
25. Bennett, J.: Obituary for John Allen Ovenstone. In: Australian Computer Bulletin (September 1984)
26. Morton, P.: ibid (1989)

27. Millar, D.D.: The Messel era: the story of the School of Physics and its Science Foundation within the University of Sydney, Australia. Pergamon Press, Sydney (1952-1987)
28. Professor Peter Aplin, Interview (2003)
29. Bennett, J.M.: Early Computer Days in Britain and Australia - some Autobiographical Snippets. In: IEEE Annals of the History of Computing 12/4 (1990)
30. Blatt, J.: News. In: The Nucleus 2/4, University of Sydney, Sydney (July 1956)
31. Basser Newsletter NS15. University of Sydney, March 14 (1966)
32. Brooks, G.: Memo to J. M. Bennett. University of Sydney, Sydney (June 6, 1968) (Archives Box 54)
33. Green, J.H., Woods, L.C.: Program for Education for Nuclear Engineering, Science at NSWUT. In: Australian Atomic Energy Symposium, NSWUT (1958)
34. Smart, R.G.: The Utecom Digital Computer. In: WRE 1957 ibid, pp. 104-1–104-5 (1957)
35. University News, UNSW, September 29 (1964)
36. Allen, M.: ADA - A Transistor Decimal Digital Differential Analyzer. In: WRE 1957 ibid, pp. 209-1–209-29 (1957)
37. Bennett, J.M., et al.: ibid, pp. 36–38 (1994)
38. Wong, D.G.: The Design and Construction of the Digital Computers Snocom, Nimbus and Arcturus. PhD Thesis, University of Sydney (1966)
39. Professor Murray Allen Interview (2003)
40. Wong, D.G.: The Investigations Leading to the Specification of a Digital Computer for Power System Operational Studies. ME Thesis, University of Sydney (1960)
41. Frankel, S.P.: The Logical Design of a Simple General Purpose Computer. Trans. IRE, PGEC EC-6(1) (March 1957)
42. Wong, D.G.: ibid (1966)
43. Bennett, J.M., Dakin, R.J.: Computers as an Aid in Computer Design Assessment. The Computer Journal 3(4) (1961)
44. Bennett, J.M., et al.: ibid, pp. 38–40 (1994)
45. ibid, pp. 44-49
46. Lavington, S.: Early British Computers. Manchester University Press (1980)
47. Kidman, B., Potts, R.: Paper tape and punched cards: the early history of computing and computing science at the University of Adelaide. University of Adelaide (1999)
48. Allen, M.W., Rose, G.A.: System Design of CIRRUS. In: First Conference on Automatic Computing and Data Processing in Australia. ANCCAC Sydney, p. C5.2 1 (1960)
49. Bennett, J.M., et al.: ibid, pp. 44–49 (1994)
50. Hinkfuss, I.C., Keith, R.J., Macaulay, I.J.: Design of a High Speed Parallel Solid State Digital Computer. In: Proc. IRE (September 1960)
51. Pearcey, T.: ibid (1988)
52. Lavington, S.: ibid (1980)
53. Hinkfuss, I.C.: A 1,200 baud Digital Data Transmission Scheme. In: Proc. IRE 1961 (1961)
54. Barlow, G.: A Real Time Data Processing System. In: Proc. NSW Comp. Soc. Conf. (September 1964)
55. ibid
56. Bennett, J.M., et al.: ibid, p. 125 (1994)
57. Wong, D.: ibid (1966)
58. Communication from John Bunton (2001)
59. Pearcey, T.: ibid (1988)
60. ibid

Why the Real Thing Is Essential for Telling Our Stories

David Demant

Senior Curator Information and Communication,
Museum Victoria, GPO Box 666, Melbourne, Victoria 3001, Australia
ddemant@museum.vic.gov.au

Abstract. Museum Victoria possesses the only intact first generation electronic stored program computer left in the world. Real things, like CSIRAC, are entry portals to a past era. Along with contemporary documents and records, they are the closest we can get to time travel. They complement historical records. Historical records are not substitutes for the real thing, neither are replicas or facsimiles. We use real objects in combination with historical and contemporary knowledge to develop our understanding of the past. The presentation answers the question implied by the title 'Why the real thing is essential for telling our stories'' in two ways. First, it discusses what we and future generations gain by conserving and interpreting the real thing on an on-going basis. Second, it gives examples from the museological work done with CSIRAC and its archive.

Keywords: first-generation; stored-program; electronic; computer; software; CSIRAC; objects; real-thing; facsimiles; replicas; museum; archive.

1 Museum Victoria

Museum Victoria possesses the only intact first generation electronic stored program computer left in the world.

It is a real thing, not a replica. It was the first computer in Australia, fourth in the world. It is complemented by a complete archive of software, documentation, paperwork and drawings – also all real things.

Real things are Rosetta stones – entry portals to a past era. Along with contemporary documents and records, they are the closest we can get to time travel. They complement historical records; they are not substitutes.

What do real things like CSIRAC and its archive share with the Rosetta Stone? The Rosetta Stone enabled us to recover information coded in Egyptian hieroglyphs. This knowledge had been lost for over 1300 years. The Stone has inscribed upon it a single text written in three different inscriptions, one of which is hieroglyphics. The representation of this text in three scripts, combined with contemporary understandings and knowledge, enabled scholars to decipher the hieroglyphs. The story of the Rosetta Stone gives us hope that 'lost' information can be recovered.

For the purposes of this paper, the story of Rosetta Stone has another important lesson. The analysis of the Stone led to the discovery that the hieroglyphic language is more than a pictorial representation of ideas; that it is a spoken language with more information embedded in it than just images. The analysis revealed more of the

A. Tatnall (Ed.): HC 2010, IFIP AICT 325, pp. 13–15, 2010.

knowledge embedded in the hieroglyphs than was expected when the analysis was initiated. In the same way, preserving the real thing may enable future generations with insights that are beyond our current state of knowledge.

We need real objects in combination with historical and contemporary knowledge to develop our understanding of the past. Our understanding of the past prepares us better to deal with the present and the future.

2 CSIRAC

Examples will be given from the work done with CSIRAC and its archive that illustrate the value of preserving the real thing compared to replicas and facsimiles. Replicas and facsimiles generally reproduce only a few aspects of an item and even those are not necessarily reproduced accurately or completely.

The work referred to was carried out by the CSIRAC History Team[1], which includes former operators and users of CSIRAC. The examples involve the extensive paper tape archive of mainly 12-hole and 5-hole punched paper tapes. This includes the CSIRAC Library tapes, which held a group of tried, tested and documented programs. There are program listings for each tape in the CSIRAC paper tape Library but other tapes were not so well blessed as regards testing and documentation.

The first example of the value of the real thing deals with the translation into electronic form of the programs punched into the original paper tapes, whether they were part of the Library or not. The equipment to do this archival work was specially designed to read the original programs (punched into the paper tape); the equipment to read paper tape was calibrated using an original manual paper tape punch, part of the CSIRAC museum collection. The existence of the original tapes and/or listing ensured accuracy of the content of the tapes. One of the incentives for this work was the desire to recreate the music of CSIRAC. CSIRAC was the first electronic computer to be programmed to generate music, in 1951.

The second example involves an accidental discovery. A further tape that was acquired fortuitously from the private collection of one of the CSIRAC pioneers,[2] provided the basis for the rewriting of an existing incomplete (and in part erroneous) program. The result was a correct and executable program. This would not have been possible without that tape. This shows that a combination of original artifacts, expertise and historical records can provide new insights.

The important thing is that these examples show that information can be resurrected rather than indicating the value of the information. The value of an original item may not be apparent initially; indeed one can never be sure what insights it may later provide.

The presentation will also include a history of CSIRAC from its development phase in Sydney through its operational phase in Melbourne to its current role as an

---

[1] The CSIRAC History Team is part of The University of Melbourne and works closely with Museum Victoria.

[2] The pioneer referred to was Dr. Frank Hirst, who headed the Computational Laboratory at the University of Melbourne. CSIRAC was transferred to The University of Melbourne in 1955 under the care of Dr Hirst. Its arrival enabled the University to establish one of the earliest Computer Science departments in the world.

icon of digital technology in Museum Victoria. It has become a symbolic milestone in the human journey.

CSIRAC is the abbreviation for Commonwealth Scientific and Industrial Research organisation Automatic Computer and is pronounced 'sigh-rack'. It was developed in Sydney and, in November 1949, it ran its first test program. In 1955, it was transported to Melbourne and, from 1956 to 1964; it provided a computing service for science and industry. In 1964, it was switched off for the final time and donated to Museum Victoria. It is currently on long-term display at Melbourne Museum.

For more information about CSIRAC, please visit:

> http://museumvictoria.com.au/csirac/
> http://www.csse.unimelb.edu.au/dept/about/csirac/

References

1. Beard, M., Pearcey, T.: The Genesis of an Early Stored-Program Computer: CSIRAC. IEEE Annals of the History of Computing (IEEE) 6(2), 106–115 (1984)
2. Deane, J.: CSIRAC: Australia's first computer, 45 p. Australian Computer Museum Society (1997), ISBN: 0-6463-4081-6
3. Demant, D.: The first computer mouse. Museum Victoria (2001), ISBN: 0731184211
4. Doornbusch, P.: The Music of CSIRAC, Australia's first computer music. Common Ground (2005), ISBN: 1-86335-569-3
5. McCann, D., Thorne, P.: The Last of The First, CSIRAC: Australias First Computer. University of Melbourne Computing Science (2000), ISBN: 0-7340-2024-4
6. Pearcey, T.: A History of Australian Computing, 192 p. Chisholm Institute of Technology (1988), ISBN: 0947186948

Wonder, Sorcery, and Technology: Contribute to the History of Medieval Robotics

Nadia Ambrosetti

Dipartimento di Informatica e Comunicazione, Università degli Studi di Milano
Via Comelico 39/41, 20135 Milano, Italy
nadia.ambrosetti@unimi.it

Abstract. The paper considers some Medieval sources about imagined or actually studied automata, to make a contribution to the reconstruction of the cultural landscape of a period that, from the technological point of view, is commonly regarded as less interesting than others. It will be shown that the idea of an automatic device or system, capable of performing not necessarily simple tasks, of measuring its own state and of taking action based on it, was well established in the Medieval mind, even though sometimes connected with magic.

Keywords: automata, design, sorcery, science, technology, feed-back control, history, magic, robot, robotics.

1 Introduction

The Greek adjective αὐτόματος, coming from the adjectives αὐτὸς (self) and ματὸς (having in mind, acting), was at first used to refer to any event, that happened spontaneously, without external intervention; later its use was extended to those mechanical devices, which perform, after a user's input, a finite number of default actions, typically, but not necessarily in a periodic sequence. Such automata were, during the Hellenism, also equipped with mechanisms for controlling their state and were so capable of taking action in dependence on the state itself (a first and essential mechanism for feedback). The purpose they were built for, was essentially playful: they were made to arouse the spectators' wonder (θαῦμα) and therefore their admiration not only for the engineer's skills, but also for the monarch, who had sponsored his designs and works. Often, these designers managed to surprise their audience through the implementation of devices activated by an input that doesn't appear to involve the produced effect: the typical example is the mechanism that opens the doors of a temple after the lighting of a ritual brazier. We rarely find examples of devices which also can have a practical use (such as a purifying water dispenser, or a pump to extinguish fires).

The most important source for this study are the works by Hero of Alexandria (1st century AD): Πνευματικὴ (Pneumatics) and Ἀυτοματοποιητικὴ (Automata building); these works came to us almost complete and they let us understand much of ancient engineering applied to automata. This fortunate tradition demonstrates that such works passed from Alexandria into the Romans' hands at first, then to the Byzantines and Arabs; during the Dark and Middle Ages they were read, understood,

A. Tatnall (Ed.): HC 2010, IFIP AICT 325, pp. 16–25, 2010.

and copied, so that today, dispersed in many European libraries, we have respectively more than 100 copies in Greek language, and 13 translated into Latin [1]. Another well-known author of a treatise about *Pneumatics* is also Philo of Byzantium [2-3]: his work, though less widespread than Hero's, was, however, read and studied during the Middle Ages and today we have 15 manuscripts of Latin translations, mostly entitled *De spiritualibus ingeniis* (About devices dealing with fluids).

2 The "Thauma Connection": From Alexandria to Baghdad, and Byzantium

When these works came into the hands of Arabic and Byzantine scholars, they were not only subject to a thorough theoretical study, but they also brought to the actual implementation of devices. While in the Byzantine empire no translation of these texts was required, in the 9th century, Qusta ibn Luqa, author of many translations of scientific works from Greek into Arabic, provided Arabic scholars with a version of Hero's *Mechanics*.

The social structure of the Byzantine and the Arabic empires was fundamentally identical to the Alexandrian one: the king (emperor or caliph) was the undisputed arbiter of cultural life; for his impressive palace, wonderful automata were constructed for the double purpose of amusing the king and his court, and of arousing the admiration of the audience, mainly ambassadors of foreign peoples.

2.1 Arabic World

In the Arabic world, the most famous scholars in this field were the brothers Banu Musa, who lived in the 9th century in Baghdad, and al-Jazari, who flourished between the 12th and 13th centuries [4-5]. Their works titles are all connected with the idea of surprising the audience, and their actual achievements were exactly in that direction, although by different paths: they tried to reproduce scenes of wildlife (birds singing on a tree), or men and women performing some actions (e.g., an orchestra of musicians playing a song on a boat floating in a pool of the royal garden; a drink-serving waitress). Some of these automata had however a practical purpose, such as a medical equipment, like the device for measuring the amount of blood drawn with the phlebotomy, or liturgical objects, such as a peacock-shaped basin for ritual ablutions.

2.2 Byzantine Empire

At the magnificent court of Byzantium, the natural heir of Hellenistic tradition, but also of the splendour of the Eastern world, automatic devices seem to have been exclusively used in the imperial Great Palace, in the Magnaura. Interestingly, in this case, the sources are not technical and no name of architect or engineer, that made automata in Byzantium, was handed down.

The emperor Constantinus VII Porphyrogenitus (913-959) composed the Ἔκθεσις τῆς βασιλείου τάξεως (Ceremonies at the Imperial Court), a compilative work. The main content is a detailed description of the ceremonies from the court's point of view; in II,15, he relates what usually happened during a typical audience granted by the emperor: after the ritual bowing, the postulant heard a roar coming from the lions on either side of the throne, while birds, resting on the trees that surrounded the throne, began to sing harmoniously. Subsequently, the lions, in perfect synchrony with the various moments of the ceremony, departed from their starting position and then returned back to it; both lions and birds gave up singing at the end of the ceremony [6-7].

A very similar ritual is also described by the bishop Liutprand of Cremona (920-972), in his *Antapodosis,* a report about his stay at the court of Byzantium as an ambassador of Berengarius II (900-966), who was eager to be accredited as king of Italy by the emperor of the Eastern Roman Empire; Liutprand was formally received by Constantinus Porphyrogenitus, in 949, and stayed for some time in the city, taking part in court ceremonies [8-9].

The presence of a tree near a king's throne [10] dates back to the Sumeric epic of Gilgamesh; a golden plate tree and a golden vine are also cited in Herodotus' *Histories* (VII, 27), as a gift to the Persian emperor Darius by the king of Lydia, Pythius, grandson of the renown Croesus [11-12]. Singing birds are likely to have been inspired by Hero's and Philo's *Pneumatics* (respectively, I, 4; I, 5; I, 16; and 61, for instance), exactly as it had happened in the Arabic world, but, given such lack of technical Byzantine sources, one could also suppose that the Byzantine engineers were inspired by Arabic designs or models possibly seen in Baghdad. The reference to such a device by the Sicilian poet Ibn Hamdis, who lived during the 11th century, could support the hypothesis of the spreading of such automaton design [13].

Concerning the lions, an iconographic source for the Byzantine engineers could have been the throne of Solomon, as described in the Bible (Kings, I, x, 18-20), though no reference is here made to any motion.

These automata show perfectly how the intellectual resources of the Byzantines (in this case, their competence in the fields of mechanical engineering) could be used as *instrumentum regni*, and they also demonstrate the effort that was made both in the imitation of nature, and in showing the emperor as a God's epiphany.

Automata descriptions within the Byzantine Empire are also present in many literary works written in order to entertain the audience: romances, and allegorical poems [14].

3 Dark and Middle Ages

After the great season of Hellenistic engineering and during all the early medieval centuries in the Eastern Roman Empire or in the territories under Arabic domination, the sources related to automata decrease dramatically and are limited in Europe to secondary references, and to romances. No designs or technical descriptions are available, at least until the 13th century.

3.1 Dark Ages

Anyway, secondary sources can testify the existence of mechanical precision devices, demonstrating that even in Roman-barbarian kingdoms the technical skills required for such objects were not completely lost.

For instance, in Cassiodorus' *Variae,* the official correspondence of the Ostrogothic court in Ravenna, a letter (XLV, 6) addressed to the philosopher Boethius is preserved; Theodoric says that the Burgundian king has repeatedly asked him for a water clock and for craftsmen (probably as maintenance men); he comments the episode with revealing words: the reason why the Burgundian king insisted so much, is that he considers as a "miraculum" (wonder) an object that for Theodoric is instead "cottidianum" (daily) [15].

Although we can suppose that Theodoric overemphasized his acquaintance with these devices, it is clear that at the beginning of the 7th century in Europe there were still craftsmen capable of building a water clock.

3.2 Middle Ages

Another important secondary source consists in romances [16]: whether they belong to the Matter of Britain (King Arthur's stories) or of France (Charles the Great's stories), or of Rome (the main character is taken from ancient history, such as Julius Caesar or Alexander the Great), references to automata are frequent; they are primarily used as guardians of a strategic place (a tomb, a bridge, a cave, or a castle) and their makers are usually magicians, not technicians.

The reason is easily explained: all mechanical arts were in no esteem throughout the Middle Ages to the point that, for a cultured man, technical practice was considered morally, religiously, and socially dishonorable [17]. On the contrary, knowledge and practice of magic were reserved for a selected group of people, who had chosen to engage in "religiously illicit pursuits, illegitimate knowledge, and trafficking with the Devil" [18]; by attributing the construction of automata to magicians, any moral conflict was avoided.

Even the legends about the French monk Gerbert d'Aurillac (946-1003), later Pope Sylvester II, and the Dominican philosopher Albertus Magnus (1206-1280) as automata builders, reveal an explicit moral intent: both episodes end with the destruction of the automaton (Albertus' talking head would have been broken by one of his students, St. Thomas Aquinas), which is moreover impossible to restore.

3.2.1 The Great Sorcerer Gerbert d'Aurillac?

In Gerbert's biography by William of Malmesbury (1080/1095-1143), an interesting event is reported [19-20]. During his stay in Rome, the future pope had decided (by sheer greed) to rescue Octavian's treasure, buried, according to a legend, in a cave beneath the Roman Forum; the cave entrance was indicated by a statue. Many others had unsuccessfully tried to find the entrance, being less artful than him: he marked the place, where the shadow of a statue's finger fell at noon-day; at night he made the earth open by means of his magical arts and, attended by a servant with a lantern, he entered a beautiful golden palace, where they found a court of golden automata. The light source was a small carbuncle of the first quality, standing on a base. At the opposite corner of the room, stood a young archer, holding a bow and an arrow. Gerbert and his servant soon noticed that, if they tried to take anything, all these automata appeared to rush forward to attack them. So Gerbert guessed what would be the consequences of such an attempt, and decided to give up, but his servant, unable to resist, decided to steal a superb knife. Immediately, the automata cried loudly and began moving, and the archer shot the carbuncle with his arrow, so that all was in darkness; Gerbert ordered his servant to drop the knife, otherwise their lives would have been in danger. He obeyed, and they left the palace, though without any treasure.

The story is very interesting from several points of view.

First of all, the palace designer had planned the system in order to kill any thieves or at least to give them a moral lesson, not to impress them with the automata's capabilities. We have the description of an actual distributed system made up of automata: they are all running the same task (stop the intruders, if they are stealing something). The actions performed by the automata are not periodical: they begin and end depending on values measured by "sensors". Nothing is said about how the system reverts to its initial state after the departure of the intruders.

Then, in Hero's *Pneumatics* (I, 41), an automaton composed by two figures, entitled *Hercules and the Snake*, is described: when a user lifts an apple (placed between the statue of the archer Hercules and the tree the snake is wrapped around) Hercules shoots with an arrow the snake, which in the meanwhile begins to hiss. Its operation is relatively simple: the base of the group is divided horizontally into two parts (the top compartment is full of water), connected by a drain hole, where a cork is set. By raising the apple, a double synchronous effect is produced:

1. a chain, connected to the apple, pulls the cork from the hole, causing the gurgling through the various ducts that is similar to a snake hiss;
2. a second chain, also connected to the apple, acts on the figure of Hercules, making him stretch and release the string of his bow.

In addition, in a French romance, whose first written version dates to the middle of the 12th century, entitled *Le pèlerinage de Charlemagne* (Charles the Great's Pilgrimage), the setting of the emperor's main adventure is the vaulted and circular palace of Hugon, emperor of Constantinople, where Charles and his 12 peers are housed as guests in a beautiful bed-chamber, full of precious decorations and refined objects. The light source is a carbuncle, set on a pillar [21]. The front of the palace is decorated with the statues of two smiling young men, holding ivory horns. Whenever a wind comes up, these images blow their horns, producing a loud clear sound, and immediately the palace begins rotating. A carbuncle and a moving room appear also in the 14th-century allegorical poem titled Σοφροσύνη (Temperance) by Theodore Meliteniotes [22]. Here we also find an abridged version of a lapidary, a book where stones are put in relation with moral qualities: the carbuncle is a symbol for temperance, a cardinal virtue, related to self-control.

Last, but not least, remark: Gerbert is told to have designed and built a water organ, during his stay in Reims; his treatise on the subject has been recently studied [23].

Even taking into consideration that this is a legendary episode, one cannot but be struck by the strong similarity between this system and the Heronian automaton, which may have been the remote source of the story, though it was considered neither in the Arabic written tradition nor in the Byzantine implementations.

It is therefore reasonable to make two assumptions, not mutually exclusive:

1. This knowledge was transmitted within the workshops by the masters to their apprentices from the Antiquity to the Middle Ages with or without Arabic contribution; it is unfortunately very hard to prove such an assumption, due to the lack of sources.
2. A Latin cultured tradition, mostly separated from the Arabic one, existed, and flourished, probably in the monasteries; this second assumption could be easier to prove, by means of a census of Latin manuscripts on the subject.

3.2.2   Looking for Pneumatics' Medieval Latin Tradition

Despite the wide diffusion of Arabic culture and technology, that had accompanied the territorial expansion of the 7th-15th centuries, at the time neither the Latin versions of Arabic engineers' masterpieces, nor information in this regard are available, since the lists of the translations carried out in Europe don't include these works. We may assume that the oldest (so far as it is known at the moment) Latin translation of Philo's *Pneumatics* (Cambridge, Pembroke College, 169) has been made from an Arabic translation: its incipit "In nomine Dei pii <et> misericordis…" is common to other Latin translations of Arabic scientific works (e.g., Robert of Chester's version of al-Khwarizmi's *al-jabr w'al-muqabalah*, or Iohannes Hispalensis' *De scientia astrolabii*) [24-25].

It is equally unlikely that Byzantine technical works existed, since in European libraries no Greek manuscripts are housed dating before the 15th century. Even the oldest copy of Hero's *Pneumatics* (Venezia, Biblioteca Marciana, gr. 516) dates back to the 13th century, but it was brought to Italy from Byzantium by Cardinal Bessarion during the 15th century.

Another part of Europe where an intense activity of translation took place, was the Norman Sicily [26]: Greek, and Arabic works were translated into Latin [27]. One of the most renowned translators in the 12th century was the archdeacon of Catania, Henricus Aristippus, whose origins and life are still largely unknown. He translated from Greek into Latin Diogenes Laertius' works, Ptolemy's *Almagest*, Plato's *Meno* and *Phaedo*, Aristotle's *Meteorologica*. In his preface to the translation of the *Phaedo*, he addresses to an Englishman who was to return home from Sicily, using the following words: " […] Habes Heroni philosophi Mechanica pre manibus, qui tam subtiliter de inani disputat […] " (you have in your hands the *Mechanics* by Hero, who so subtly deals with the void). These words led Rose and Birkenmajer [28-29] to believe that it was the *Pneumatics*, since in the *Mechanics* Hero doesn't speak of void; Haskins [30-32] refused such an inference, because the content of *Mechanics* is known only from Arabic translations, possibly incomplete; moreover, Haskins rejected the subsequent inference by the above mentioned scholars as a risky assumption: according to them, Henricus was referring to his own Latin translation, since other translations of his are quoted in the passage.

Birkenmajer [33] goes further than Rose, stating that he had found evidence of a copy of this translation in the *Biblionomia* (the library catalogue of the French philosopher and poet Richard de Fournival, living in the 13th century), where the *Pneumatics* would be referred to as *Excerpta de libro Heronis de specialibus ingeniis* (excerpts from Hero's book about special devices); "specialibus" is probably a wrong transcription of "spiritualibus" (concerning with void), due to the fact that the shortening of the two words is identical. Unfortunately, though the library is now part of the Bibliothèque Nationale de France, no evidence of the manuscript is now available, and even the assumptions that it could be Philo's instead of Hero's *Pneumatics*, or a Latin translation of an Arabic compilation from Banu Musa's work, can be neither validated nor rejected [34].

Birkenmajer also argues that, in the 13th century, the Flemish Dominican William of Moerbecke would have translated, among many other works, Hero's *Pneumatics* into Latin; his proof is based on the presence of a treatise entitled *De aquarum conductibus et ingeniis erigendis* in a list of works, owned by St.Thomas, actually translated by the Flemish scholar [29]. Although the evidence is not completely convincing, it is remarkable that the Aquinas, already present in the legend about Albertus Magnus' automaton as a minor character, returns as a possible reader of a Hero's text.

Another work, entitled *De inani et vacuo* (About void), is both quoted without any reference to the author by the 14th-century philosopher Marsilius of Inghen in his *Quaestiones super VIII physicorum libros* (Questions on the Eight Books of the Physics; IV, 13), and copied in 1466 in a manuscript (Krakow, Biblioteka Jagiellonska, MS 568, ff. 207-211), where it is attributed to Hero. The Polish manuscript has two interesting issues:

- in the colophon the scribe's name of the antigraph is mentioned (*quem inscripsit Landfridus*); such a name (in the German version Lantfrid) is present also in the colophon of a Carolingian manuscript (München, Bayerische Staatsbibliothek, Clm. 14461, f.150) of religious content, copied in the 820s in Freising, and housed in St.Emmeram's abbey library, near Regensburg, at least since 1347. If the scribe would be the same, we would have an evidence of the connection between a Pneumatics tradition in Latin language and a Benedictine monastery, a place quite advanced from the technological point of view during the Middle Ages;
- the handwriting is very similar to Regiomontanus', the famous German astronomer and mathematician; in addition, we must say that the manuscript was bound in Germany, and the author of the letters included in the last part of the manuscript was a friend of Regiomontanus': Cristianus Roder de Hamburgo. Regiomontanus is said by the French humanist Pierre de la Ramée to have built an eagle, and a fly, that could fly away and back. Despite the fact that the story told by Pierre is certainly exaggerated, the fact that Regiomontanus may have dealt with an abridged version of *Pneumatics*, makes it less improbable.

Though there is no definitive proof of direct knowledge of *Pneumatics* during the Middle Ages, we, however, have 28 copies of Latin translations from Philo's or Hero's treatises made in the 14th-15th century. Based on a comparison of the incipits, they can be grouped into at very least 3 families:

- *Quum/Cum apud antiquos*: 12 manuscripts.
- *Quoniam tuum*, where the incipit mentioned above (*In nomine Dei* ...) is often, but not always, placed before: 15 manuscripts.
- *Cum spirituale negocium*: 1 manuscript.

A deep comparative study of all these manuscripts is in progress.

3.2.3 A Useful Mechanical Automaton Called Maurizio

Concerning craft tradition, we can remark that more or less in the same years one of the oldest mechanical automata appeared in Italy, in Orvieto. In 1347 the Opera del Duomo (Committee on the cathedral works) charged the clockmaker Francesco with building a clock mechanism, which required 285 pounds of iron, three blacksmiths, and eight craftsmen. The following year a bronze automaton was added to the clock as a striking system; it represents a "dottiere", a yard overseer, who was in charge of verifying compliance with working hours, and of forbidding workers to waste time.

This is one of the rarest, and oldest, instances of automata (before the Renaissance), due inter alia to a virtually unknown clockmaker. In addition, this is clearly a device which was rather designed for practical purposes than to impress an audience: the automaton had become so familiar to the citizens of Orvieto, and to the yard workers, that it was even given a name: Maurizio (distortion of the word "muricçio", that indicates the yard).

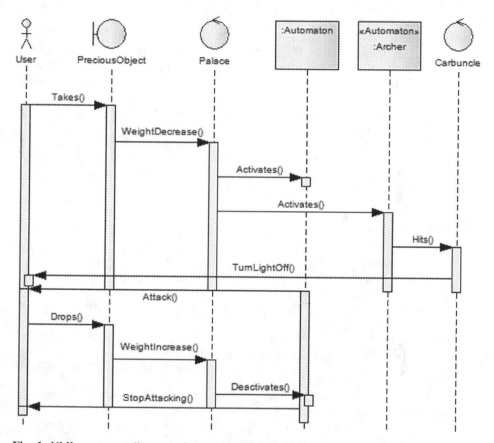

Fig. 1. UML sequence diagram of the episode of Gerbert d'Aurillac and Octavian's palace, from William of Malmesbury's biography

4 Conclusions

As it is clear from the above, the Medieval landscape is fragmented and often uncertain, and it must be frequently reconstructed upon secondary sources, such as literary texts, or hearsay. All sources and cross references need to be verified by means of careful philological and/or historical work.

In addition, such a thorough collection of sources would be useful only for documentary purposes, and would make it impossible to fully understand similarities and differences in implementation among automata designed in different ages. To solve this problem and not to reduce their works to a mere display of erudition, some scholars [16, 18, 35-37] have already attempted a classification of automata, based upon different features: their being connected with musical devices, rather than stand-alone; their being human figures rather than objects, or animals; the role they play in the story, as guardians, or entertainers; the finesse of their movements.

These classifications, though useful, cannot, however, fully capture the variety of behavior of automata in the same class, or any similarities in behavior among automata belonging to different classes. Consequently, it may be useful to introduce a different approach, based on a standard, language-independent representation of the automata's behavior.

To this purpose, at an advanced stage of sources' collection, it would be useful to make a modeling of the automata's behavior with UML diagrams; we believe that the diagram that can better represent the temporal sequence of actions and interaction between user and any automaton or system of automata is the sequence diagram. An example of this representation is given below, in relation to the above narrated episode of Gerbert d'Aurillac.

The next research step will therefore be to implement a database in which diagrams, similar to the above, will be added to quotations from the corresponding passages of literary and/or technical works, in order to allow abstraction and analysis of persistent archetypal elements, such as ideas and designs.

References

1. Schmidt, W., Nix, L.L.M., Schöne, H., Heiberg, J.L. (eds.): Heronis opera quae supersunt omnia. Teubner, Leipzig (1899)
2. Prager, F.D. (ed.): Philo of Byzantium, Pneumatica: the first treatise on experimental physics. Ludwig Reichert Verlag, Wiesbaden (1974)
3. Drachmann, A.G.: Ktesibios, Philon and Heron: A Study in Ancient Pneumatics, Munksgaard, Copenhagen (1948)
4. Hill, D.R.: The Book of Ingenious Devices: Kitáb al-Hiyal. By The Banú (sons of) Músà bin Shákir. Reidel Publishing Company, Dordrecht (1979)
5. Hill, D.R. (ed.): The Book of Knowledge of Ingenious Mechanical Devices by Ibn al-Razzaz al-Jazari. Springer Science, Dordrecht (1974)
6. Reiske, J.J. (ed.): Constantini Porphyrogeniti Imperatoris De Ceremoniis Aulæ Byzantinæ. Weber, Bonn (1829)
7. Lewis, M.J.T.: Antique Engineering in the Byzantine World. In: Lavan, L., Zanini, E., Sarantis, A. (eds.) Technology in Transition A.D. 300-650, Brill, Leiden, pp. 367–378 (2007)
8. Ravegnani, G.: L'ambasceria di Liutprando di Cremona alla corte di Costantino Porfirogenito e il Libro delle cerimonie. In: SYNDESMOS, pp. 323–337. Università di Catania, Catania (1994)
9. Muratori, L.A. (ed.): Rerum Italicarum Scriptores, Ex typographia Petri Cajetani Viviani, Florentiae, vol. II, caput II, pp. 469–470 (1748)
10. Jacobsthal, P.: Ornamente Griechischer Vasen - Aufnahmen, Beschreibungen und Untersuchungen. Frankfurter Verlags-anstalt, Berlin (1927)
11. Briant, P.: Histoire de l'Empire perse. Librairie Artheme Fayard, Paris (1996)
12. Herodote: Histoires. Tome VII: Livre VII: Polymnie. Texte établi et traduit par Ph.-E. Legrand. Les Belles Lettres, Paris (1986)
13. Ruta, C. (ed.): Poeti arabi di Sicilia. Edi.bi.si, Messina (2009)
14. Trapp, E.: Learned and Vernacular Literature in Byzantium: Dichotomy or Symbiosis? Dumbarton Oaks Papers 47, 115–129 (1993)

15. Mommsen, T., Traube, L. (eds.): Cassiodori senatoris Variae. apud Weidmannos, Berolini (1894)
16. Bruce, J.D.: Human Automata in Classical Tradition and Mediaeval Romance. Modern Philology 10, 511–526 (1913)
17. White Jr., L.T.: Medieval Engineering and the Sociology of Knowledge. The Pacific Historical Review 44, 1–21 (1975)
18. Truitt, E.R.: Trei poëte, sages dotors, qui mout sorent di nigromance: Knowledge and Automata in Twelfth-Century French Literature. Configurations: A Journal of Literature, Science, and Technology 12, 167 (2004)
19. Sharpe, J. (ed.): William of Malmesbury's Chronicle of the Kings of England. Henry G. Bohn, London (1847)
20. Migne, J.P. (ed.): William of Malmesbury De gestis regum Anglorum libri quinque. Patrologia cursus completes, Series Latina, Paris, vol. II, pp. 179–181 (1862)
21. Legrand, E.: Bibliothèque grecque vulgaire. Mainsonneuve, Paris (1880)
22. Miller, E.: Poème allégorique de Meliténiote. Notices et extraits des manuscrits de la bibliothèque impériale et autres bibliothèques 19, 1–138 (1858)
23. Flusche, A.M.: The Life and Legend of Gerbert of Aurillac: The Organbuilder Who Became Pope Sylvester II. Edwin Mellen Press, Lewiston (2006)
24. van Egmond, W.: Practical Mathematics in the Italian Renaissance: A catalogue of Italian Abbacus Manuscripts and Printed Books to 1600. Istituto e Museo di Storia della Scienza, Firenze (1981)
25. Ambrosetti, N.: L'eredità arabo-islamica nelle scienze e nelle arti del calcolo dell'Europa medievale. LED edizioni, Milano (2008)
26. Houben, H., von Sizilien II, R.: Herrscher zwischen Orient und Okzident. Wissenschaftliche Buchgesellschaft, Darmstadt (1997)
27. Setton, K.M.: The Byzantine Background to the Italian Renaissance. Proceedings of the American Philosophical Society 100, 1–76 (1956)
28. Rose, V.: Die Lücke im Diogenes Laertius und der alten Übersetzer. Hermes: Zeitschrift für klassische Philologie I, 373–385 (1866)
29. Birkenmajer, A.: Vermischte Untersuchungen zur Geschichte der mittelalterlichen Philosophie. Aschendorffschen, Munster (1922)
30. Haskins, C.H.: Further Notes on Sicilian Translations of the Twelfth Century. Harvard Studies in Classical Philology 23, 155–166 (1912)
31. Haskins, C.H.: Studies in the History of Medieval Science. Harvard University Press, Cambridge (1924)
32. Haskins, C.H.: The Sicilian Translators of the Twelfth Century and the First Latin Version of Ptolemy's Almagest. Harvard Studies in Classical Philology XXI, 75–102 (1910)
33. Birkenmajer, A.: La Bibliothèque de Richard de Fournival. In: d'Alverny, M.T. (ed.) Études d'histoire des sciences et de la philosophie au Moyen Age. Studia Copernicana, vol. 1, pp. 117–210. Zakład Narodowy im. Ossolińskich, Wrocław (1970)
34. Grant, E.: Henricus Aristippus, William of Moerbeke and Two Alleged Mediaeval Translations of Hero's Pneumatica. Speculum 46, 656–669 (1971)
35. Pugliara, M.: Il mirabile e l'artificio: creature animate e semoventi nel mito e nella tecnica degli antichi. L'erma di Bretschneider, Roma (2003)
36. Söhring, O.: Werke bildender Kunst in altfranzösischen Epen. Romanische Forschungen 12, 493–640 (1900)
37. Faral, E.: Le merveilleux et ses sources dans les descriptions des romans français du XIIe siècle. In: Faral, E. (ed.) Recherches sur les sources latines des contes et romans courtois du Moyen Age, Champion, Paris, pp. 307–388 (1913)

Andrew D. Booth – Britain's Other "Fourth Man"

Roger G. Johnson

Dept of Computer Science and Information Systems
Birkbeck University of London
Malet Street, London WC1E 7HX, UK
r.johnson@bcs.org.uk

Abstract. Andrew Donald Booth (1918-2009) was the leader of a team of computer pioneers at Birkbeck College in the University of London, UK. Booth worked with limited resources, both human and financial, and concentrated on building smaller machines. This paper presents an outline of his career in the UK which, the author believes, has not received the attention it deserves in comparison to a number of his UK contemporaries.

Keywords: Andrew Booth, Kathleen Booth, Norman Kitz, computer pioneer, Booth multiplier, magnetic drum, natural language translation, desktop computer.

1   Introduction

Andrew Donald Booth, (born Feb 11th 1918 and died Nov 29th 2009), was the leader of a team of computer pioneers at Birkbeck College in the University of London. This paper presents an outline of his career in the UK (he left for Canada in 1962) which, the author believes, has not received the attention it deserves in comparison to a number of his UK contemporaries.

Following the end of World War 2, four groups in the UK were looking at building digital computers. The best known were at Manchester (associated with Freddie Williams and Tom Kilburn) and at Cambridge (led by Maurice Wilkes). The third group was at the National Physical Laboratory, Teddington (led by Jim Wilkinson and Ted Newman using a design by Alan Turing). The fourth group was that of Andrew Booth at Birkbeck College, London University. He worked with limited resources, both human and financial, and concentrated on building smaller machines. He had a radical ambition for that time of building a computer that was cheap enough that each university could own one! This was at a time when the NPL ACE was being talked of (at least at NPL) as sufficient for the whole of the UK's needs!

This paper highlights Booth's unusual combination of skills – a first class mathematical mind capable of working on the complexities needed to handle the mathematics of X-ray crystallography for a PhD combined with a natural gift for practical engineering which enabled him to design and build innovative relay and electronic computers.

A. Tatnall (Ed.): HC 2010, IFIP AICT 325, pp. 26–37, 2010.

2 Relay Computers

Andrew Booth first met the challenge of solving complex sets of equations while working on the X-ray structure of explosives during World War 2. In 1975 in an interview with Christopher Evans for the Science Museum, London [1] he related how during the war he managed a small team of girls doing these calculations and being by temperament a mathematician I don't like arithmetic. .... I didn't think much of the methods they were using and I tried to do two things. In the first place. I devised some better mathematical methods ... but I also made one or two small hand calculators.

His father was a marine engineer and part time inventor. Consequently it was perhaps not surprising that Andrew Booth developed mechanical devices to reduce the need for laborious calculation by hand of solutions to sets of equations for their many observations.

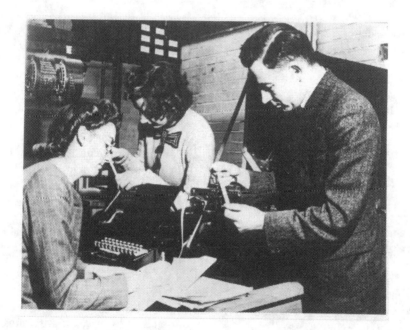

Fig. 1. Kathleen Britten, Xenia Sweeting and Andrew Booth working on ARC in December 1946

It was this latter interest that brought him to the attention of the great crystallographer J D Bernal. Returning to Birkbeck College in the University of London at the end of WW2, Bernal started building a new research group to study crystallography. He decided to appoint four assistants, one of whom was to lead on mathematical methods. He appointed Andrew Booth who had completed a PhD on crystal structures of explosives at Birmingham in 1944. Andrew Booth started by building analogue devices and exploiting other mechanical devices as he outlined in his first book [2].

Shortly after his arrival at Birkbeck he started to build his first electromechanical calculator, the Automatic Relay Calculator (ARC), shown in Figure 1. Due to a lack of space at Birkbeck the calculator was built at the British Rubber Producers Research Association in Welwyn Garden City where he had been briefly employed between Birmingham University and Birkbeck.

In 1946 Bernal obtained funding from the Rockefeller Foundation for Andrew Booth to visit US researchers working on computers. Andrew Booth reported that only von Neumann (a friend of Bernal) at Princeton gave him any time. In 1947 Andrew Booth undertook a 6 month US tour, again funded by the Rockefeller Foundation, based at the Institute of Advanced Studies at Princeton with John von Neumann and accompanied by his research assistant, Kathleen H V Britten, who was soon to become his wife.

3 Building a Memory

One major result of the 1947 visit was the redesign of the original ARC to give it a "von Neumann" architecture (the resulting design being usually referred to as ARC2). Andrew and Kathleen Booth set out the technological options for each of the components of a computer using a "von Neumann" architecture in a paper which circulated among the growing community of computer pioneers. Such was the interest, they produced a second edition [3].

The heart of the "von Neumann" architecture was the memory. In the paper they evaluated all of the physical properties including heat, light, sound and magnetism and concluded that magnetism offered the best prospects because of its persistence.

Andrew Booth was interested in building a low cost computer and so needed low cost components. On his trip around the USA he had seen a simple recording device sold for use in commercial offices which allowed managers to record letters on to magnetic oxide coated paper discs for typing by their secretarial staff. However, in order to achieve the performance needed to act as the memory of a computer he had to rotate the paper disc much faster than for simple voice recording. At this higher speeds it proved impossible to keep the disc flat and so he had to abandon this first attempt at a floppy disc.

Andrew Booth's next attempt used a metal drum. The first drum was mounted horizontally and about the size of a cotton reel, being 2 inches in diameter with a modest packing density of just 10 bits per inch. The drum was made of brass with a nickel coating. Thus it was that Andrew Booth built the world's first rotating electronic storage device albeit a drum rather than the now ubiquitous disc. This drum, shown in Figure 2, is now on display in the Science Museum, London.

The drums were built by his father and together they created a small company called Wharf Engineering Ltd which manufactured small discs and other computer peripherals.

During the 1947 visit Andrew Booth met Warren Weaver, Natural Sciences Division Director of the Rockefeller Foundation, who had funded the trip. Andrew Booth asked if the Foundation would fund a computer for London University. Weaver said that the Foundation could not fund a computer for mathematical calculations but that he had begun to think about using a computer to carry out natural language translation and that the Foundation probably could fund a computer for research in that area.

Fig. 2. Booth's First Drum   Photo Courtesy of NMSI. London

These events gave rise to the first official reference to computing at Birkbeck, in the 1947-8 College Annual Report, which says:

An ambitious scheme is in progress for the construction of an Electronic Computer, which will serve the needs of crystallographic research at 21-22 Torrington Square; it will also provide a means of relieving many other fields of research in Chemistry and Physics of the almost crushing weight of arithmetic work, which they involve. [4]

Notwithstanding the Rockefeller Foundation's funding of a machine expressly for natural language processing, readers will note the College's emphasis on the "un-funded" mathematical calculations although, in fairness, Birkbeck became for the next fifteen years a leading centre for natural language research. Initially the tiny memory on computers meant it was very difficult to do any serious natural language processing but Andrew Booth and his research students developed techniques for parsing text and also for building dictionaries. Numerous papers and several books were published as a result and this work is discussed further in section 6.

Andrew Booth, even in the days of cumbersome early machines, wrote about making computers available as widely as possible. Nonetheless the following extract from the College report for 1949-50 under the unpromising heading of "Desk Calculating Machines" seems well ahead of its time:

The Committee of the Privy Council for Scientific and Industrial Research have made a grant for a programme of research on desk calculating machines to be carried out over the next two to three years on behalf of the National Physical Laboratory by the Electronic Computer Laboratory at Torrington Square. [5]

This project seems to have ended prematurely without a full prototype being built but a copy of Andrew Booth's report from 1950 has recently been found in the Science

Museum archives [6]. In it Andrew Booth evaluates the technical options for putting computers on, if not the desktop, at least the laboratory bench. The design used dekatron valves which operated on a decimal basis and thus provided simple counting devices. Booth lacked any simple way to display digits electronically and proposed to use the position of the lighted cathode as a counter in a manner similar to reading an analogue clockface. Andrew Booth observes in the report that this feature had been regarded as a serious shortcoming by reviewers from the National Physical labora-tory although Booth predicts correctly that a technical solution would soon become available.

Around 1948/49, Andrew Booth redesigned the ARC2 as an entirely electronic machine which he called Simple Electronic Computer (SEC). This was built by Norman Kitz (formerly Norbert Kitz), see Figure 3, and is written up in his 1951 MSc (Eng) dissertation [7].

Fig. 3. Norman Kitz working on SEC, December 1949. The larger drum in the forefront of the picture is now in the Science Museum, London.

An interesting historical footnote is that Norman Kitz left Birkbeck to work for English Electric at NPL on the DEUCE computer. From there he moved to Bell Punch and designed the world's first electronic desktop calculator, called ANITA. So although Andrew Booth never completed a desktop calculator at Birkbeck, it seems likely that he inspired one of his students to do so.

Andrew Booth moved swiftly on to create his best known computers, the All-Purpose Electronic Computers (APEC). The 1951/52 College Annual Report proudly records that *"The APEXC calculator operated successfully for the first time on 2nd May 1952"*. A year later the 1952/3 College Annual Report records that:

Fig. 4. APE(X)C Computer in 1956

The second digital computer APEXC [SEC was the first] has been completed and is in use. It has shown the expected speed of about several hundred times as fast as mechanical methods but has exceeded expectation in its reliability and freedom from breakdown. An improved model is almost complete and will take its place as soon as the first is sent to its owners, the British Rayon Research Association. [8]

The Annual Reports are slightly confused with regard to the computers' names. It is possible the names changed over time. Andrew Booth subsequently refers to the Rayon Research machine as APE(R)C – R for Rayon - and the later APE(X)C – X for X-ray - was the Birkbeck crystallographers machine, see Figure 4.

4 Booth Multiplier

If the drum reflected Booth's engineering talent, then the Booth multiplier was a demonstration of his mathematical skill. A key component of any computer design is the arithmetic unit and to provide fast arithmetic it is necessary to have hardware multiplication and division. When Booth visited von Neumann in 1947 he obtained details of von Neumann's design for both a hardware multiplier and divider. Booth described them in his interview with Evans as "a beautiful divider" but the multiplier as "an abortion" [1]. When Booth asked von Neumann why he had not used a similar approach in his multiplier as in the divider, von Neumann assured him it was a theoretical impossibility and Booth accepted the great man's opinion. Booth told Evans that when he was designing the APEC computer he realised that von Neumann was wrong and Booth recollected how, over tea with his wife in a central London cafe, he designed a non-restoring binary multiplier which, with a subsequent minor modification by a colleague, is the Booth multiplier which is still in use today.

Basically the Booth multiplier follows the usual method for long multiplication of summing partial products. However it also uses a "trick" that to multiply by a string of 9s it is possible to left shift an appropriate number of places and subtract the multiplier from the result. This approach works even better in binary where it results in a simple rule:

Examine each pair of digits in the multiplier creating the first pair by appending a dummy 0 at the least significant end, then

```
If the pair is 01, add the multiplicand
If the pair is 10, subtract the multiplicand
Otherwise, do nothing
Shift both the partial product and multiplier one
     place right and examine the next pair of digits
Repeat as many times as there are digits in the
     multiplier.
```

This was submitted for publication in August 1950 and published the following year [9].

5 Commercial Success

Accommodation at bomb damaged Birkbeck was still at a premium throughout this period and so Andrew Booth built his APE(R)C and probably later computers in a barn in Fenny Compton, Warwickshire where his father lived.

It was to this barn in March 1951 that a three man team led by Dr Raymond "Dickie" Bird from British Tabulating Machines (BTM) came to visit. BTM were the UK's leading supplier of punch card systems and their management had decided that they needed a small computer to improve the calculating power and flexibility of their tabulators.

At the time that BTM joined forces with Andrew Booth there were, as already noted, three other electronic computer projects in the UK. However, strong links had developed between the EDSAC team at Cambridge and Lyons who were building their LEO (Lyons Electronic Office) computer. Manchester were forging links with Ferranti, who like NPL with Pilot Automatic Computing Engine (ACE), were building large and expensive scientific computers.

In just a few days Raymond Bird's team had copied Andrew Booth's circuitry. Returning to BTM's factory at Letchworth they added extra I/O interfaces and named the resulting computer the Hollerith Electronic Computer (HEC), see Figure 5. This prototype computer is one of the world's earliest surviving electronic computers, unlike so many early machines which were dismantled when no longer needed, and is now in store in the Birmingham Museum.

BTM moved ahead rapidly getting HEC1 to work by the end of 1951. BTM management decreed that the HEC would go to the Business Efficiency Exhibition in October 1953 and so a new machine (HEC2) had to be built contained in a smart metal cabinet suitable for the public to see. Eight similar machines were sold as the HEC2M mainly for technical applications. The successor was the HEC4 which was a commercial data processing machine of which over 70 were sold in the UK and abroad. At the

Fig. 5. BTM HEC 1 Prototype in store at the Birmingham Museum

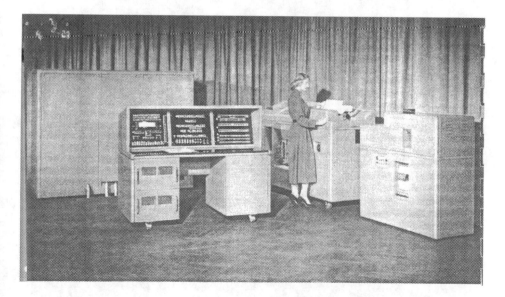

Fig. 6. ICT 1201 of 1956

end of the 1950s this was the UK's best selling computer by volume. After BTM merged with Powers SAMAS to form ICT the HEC4 became the ICT 1200 range, see Figure 6. The technical details of the HEC4 form the majority of Raymond Bird's thesis [10].

Booth continued to build new machines. After the APEC machines, came MAC (Magnetic Automatic Calculator). Three examples of a development of MAC named M.2 were built by Wharf Engineering Ltd. These were for University College London, Kings College London and Imperial College London. The Annual Report for 1957/58 notes.

The keynote of the M.2 is, as in previous machines, small size and simplicity, and an idea of what has been achieved is provided by the fact that M.2 occupies a space rather less than that of an office desk, consumes as much power as an electric fire, but has roughly the speed and capacity of the much larger commercial machines which are being provided for some of the smaller Universities. [11]

6 Natural Language Processing

The Booths with their research students published numerous books and papers on text processing including creating Braille output and natural language translation. A detailed assessment of Booth's early work in this area can be found in a paper by Hutchins [12]. On November 11th 1955 the laboratory gave an early public demonstration of natural language machine translation, see Figure 7.

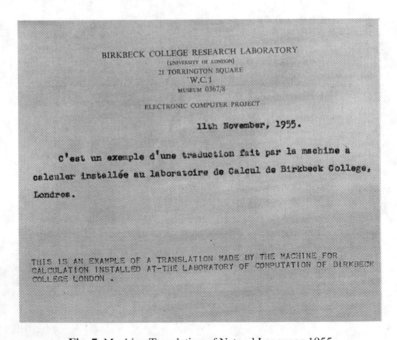

Fig. 7. Machine Translation of Natural Language 1955

The 1956/57 College Report reported that language processing work continued apace:

> Work on machine translation of languages has, however, continued at full speed and with the completion of the French programme for scientific texts attention has been transferred to German and to aspects of Platonic chronology. The laboratory was fortunate in having Dr Rais Ahmed, from the University of Aligarh, as a guest during part of his sabbatical leave and his presence gave considerable impact to the work on spoken word recognition. [13]

One of Birkbeck College's best known former students and now a College Fellow, Dame Steve Shirley who later founded a major UK software house that became Xansa, remembers as a student of mathematics and computing at that time:

> intoning "one, one, one .... two, two, two ...." into a tape recorder for some very early voice recognition research led by Andrew Booth [14]

Their wider contribution to this community has been assessed in the book by John Hutchins in his book *Early Years in Machine Translation* [15].

7 Other Achievements

Andrew Booth was an innovator in many areas of his life. The College Annual Report for the 1956/57 records the Governors' resolution on July 18th 1957 that:

From the beginning of the next session [October 1957] the Computer Laboratory be constituted as a separate Department under the Headship of Dr A D Booth [13]

The author has been told that the Department of Numerical Automation was the first academic department established to teach computing in a UK university and possibly worldwide, elsewhere the courses were still taught in computer laboratories. Also the department's M.Sc in Numerical Automation started in October 1957 was the first degree programme, many others, including at Birkbeck, having been Postgraduate Diplomas. These are difficult claims to verify and the author would welcome any information of earlier activity.

Outside the Department, Andrew Booth's played a key role through his appointment as chairman of a committee formed to set up a "National Computer Society". The 1956/57 Report notes that when it was formed in June 1957 he was elected to serve on the first Council of the British Computer Society, [13].

One notable landmark was Kathleen Booth's book on programming the APEC computers [16]. This was an early book on programming and unusual in having a female author. She did most of the programming while Andrew Booth built the computers.

The College Report of 1958/59, in a foretaste of much more recent work, reports Dr Kathleen Booth developing a program to simulate a neural network to investigate ways in which animals recognise patterns. The following year reports a neural network for character recognition.

The College Report for 1961/62 recorded that Andrew and Kathleen Booth resigned at the end of the 1961/62 academic year. Andrew Booth has given his account of the circumstances in two places [1, 17] attributing it to the College not conferring a Chair on him. In retrospect this was a massive loss to the College and, from today's perspective, totally incomprehensible given his key contributions to computer technology and his substantial research output. Andrew Booth moved to Canada where he continued his distinguished academic career initially at the University of Saskatchewan and subsequently as President of Lakehead University, Ontario.

8 Conclusions

There can be little doubt that Andrew Booth made a major contribution to the early development of computing. His two main technological contributions are the drum store and the Booth multiplier. A modified version of his multiplier is still used today in many processors. The design of his APE(X)C computer was used to provide the key components of the highly successful ICT 1200 computers, primarily used for commercial data processing. He built and sold low cost, smaller computers mainly for academic use. His M.2 series, built by his own company Wharf Engineering Ltd, were used successfully to teach programming in several London university colleges.

In the area of applications he devoted much time to the processing of X-ray crystallography data. As his academic independence increased he and his wife contributed to a number of fields notably the early development of natural language processing which is acknowledged in that community.

Andrew Booth is largely unknown outside the specialised world of the computer historian but the author believes that he deserves greater recognition by anyone whose computer writes to their hard disk or executes a multiplication instruction.

References

1. Evans, C.: Pioneers of Computing. Audio interview with A D Booth, Pioneers of Computing No 9. In: Science, Museum, London (1975) (unpublished)
2. Booth, A.D.: Fourier Technique in X-ray Organic Structure Analysis. Cambridge University Press, Cambridge (1948)
3. Booth, A.D., Britten, K.H.V.: General Considerations in the Design of an All Purpose Electronic Digital Computer, 2nd edn. Birkbeck College, London (1947) (unpublished Report)
4. Birkbeck College, London. Annual Report for August 1947, pp. 1-2
5. Birkbeck College, London. Annual Report for September 1948, p. 7
6. Booth, A.D.: DEC – Desk Electronic Computer, Science Museum Archives (1950) (unpublished report to DSIR)
7. Kitz, N.: A Discussion of Automatic Digital High-Speed Calculating Machines with special reference to SEC – A Simple Electronic Computer. MSc (Eng.) thesis. Birkbeck College, London (1951)
8. Birkbeck College, London. Annual Report for March 1952, p. 22
9. Shirley, S.: A Signed Binary Multiplication Technique. Quarterly Journal of Mechanical and Applied Mathematics 4(Pt2), 236–240 (1951)

10. Bird, R.: The Design of an Electronic Digital Computing System, PhD thesis, Birkbeck College, London (1957)
11. Birkbeck College, London. Annual Report for August 1957, p. 31
12. Hutchins, W.J.: From first conception to first demonstration: the nascent years of machine translation, 1947-1954. Machine Translation 12(3), 195–252 (1997)
13. Birkbeck College, London. Annual Report for July 1956, p. 26
14. Shirley, D.S.: Private communication with author (2007)
15. Hutchins, W.J.: Early Years in Machine Translation. John Benjamin Publishing Company (2000)
16. Booth, K.H.V.: Programming for an Automatic Digital Calculator, Butterworths, London (1958)
17. Booth, A.D., Booth, A.D.: An Autobiographical Sketch. Annals of the History of Computing 19(4), 57–63 (1997)

The Many Dimensions of Kristen Nygaard, Creator of Object-Oriented Programming and the Scandinavian School of System Development

Drude Berntsen[1], Knut Elgsaas[2], and Håvard Hegna[3]

[1] Director at the Norwegian Computing Center 1970-1989
drudeb@broadpark.no
[2] Researcher/Project Director at the Norwegian Computing Center 1963-1973
elgsaas@c2i.net
[3] Senior Research Scientist at the Norwegian Computing Center 1962-2006
hegna@nr.no

Abstract. Professor Kristen Nygaard (1926-2002) was a multi-talented scientist whose achievements were amply honoured in his later years. For general readers of Computer Science history Nygaard will be recognized as the creator with Professor Ole-Johan Dahl of the SIMULA programming languages and object-oriented programming. But Nygaard also had a social and political engagement that over the years brought him from a liberal-democratic to a socialist position where solidarity and empowerment were the major chords. This paper gives a condensed descriptive tour of the life of this many-sided computer scientist with an emphasis on how the scientific and political sides of Nygaard worked hand in hand, leading to his active cooperation with the trade unions and making him a strong advocator of the Scandinavian School of System Development and Participatory Design.

Keywords: Kristen Nygaard, computer history, object-oriented programming, operations research, participatory design, The Norwegian Computing Center.

1 Introduction

Many people are characterized as great. What constitutes great people? There is no simple answer. How do you measure greatness? Was Professor Kristen Nygaard  (1926-2002) a great man? Maybe not, but in our view Nygaard set a high standard against which you can measure other people's greatness.

Alan C. Kay, the creator of the Smalltalk programming language, once expressed that *"Kristen Nygaard is a guy who is larger than life in almost any possible dimension"* [10]. Strong words, too strong one would say if judging Nygaard's place in information processing only by his ground-breaking work with Professor Ole-Johan Dahl to establish the SIMULA language and object-oriented programming.

Kristen Nygaard in the mid-1990s

A. Tatnall (Ed.): HC 2010, IFIP AICT 325, pp. 38–49, 2010.

In this paper we will also present some of the other dimensions of Nygaard, his political work, his social engagement and cooperation with the trade unions, and his international entrepreneurship. And we will try to show that he continually worked to establish object-orientation not only as a programming tool, but as a way of organizing and communicating about system descriptions from different perspectives and as a fundamental tool in the system development process.

2 Early Years

According to his eldest son Marius [23], Nygaard showed exceptional and diverse talents from an early age. Initially his main interests were in the natural sciences and in mathematics. While he was in secondary school (age 14-15), this brought him into discussions of botany and astronomy with university professors. He followed university-level lectures and won a national mathematics award before finishing grammar school (age 16-19) [15].

His insistence on knowing everything about the topics that caught his interest was well-known among his friends and colleagues. During his years as a student, his broad music interest paid off by giving him extra financing from writing reviews of recordings of classical music in a leading Norwegian newspaper. In his later years, he took a deep interest in photography and developed expert knowledge about good wine. Nygaard's many talents and interests were aided by an immense memory capacity[1].

3 Early Professional Years

Nygaard's studies at the University of Oslo led to bachelor degrees in astronomy and physics and to a master in mathematics in 1956 based on his thesis "Theoretical Aspects of Monte Carlo Methods". The thesis theme resulted from his work at The Norwegian Defence Research Establishment (NDRE) where he had been working full time since he started his military service in 1948. His first six years at NDRE were spent in the Mathematics Section, initially doing Numerical Analysis and Computer Programming.[2] It was here that he first met Ole Johan Dahl.

In 1952 he was asked to join the Operations Research (OR) Group at NDRE, soon he became the head of the group. One typical military OR project at the time was the study of the combat capacity of infantry soldiers. How much could they carry for how long on an outdoor march over several days in rough terrain, how many calories were necessary, and what should they eat and drink, to be able to meet an enemy after several days of strain? One factor is often left out of such experiments, the willingness of the guinea pig recruits to take the experiment seriously and play the situation correctly. Nygaard and his group chose to join the experiment, do the march, eat the

---

[1] As an example, during a breakdown after a particularly hardworking period in 1962-63, which surpassed even his extreme working capacity, he was ordered by his doctor to rest and do something completely different for a while. Nygaard did rest, by reading the 22 volume world history by Swedish historian Carl Grimberg. Thirty years later he could without problems recall world history in impressive detail, and use it creatively in his political strategies.
[2] The section was headed by Jan V. Garwick (1916-89), the inspiring father of Norwegian computer science and Norway's first member of the IFIP Council [2].

food, and carry the burden just like the recruits themselves. This was an early example of action research on Nygaard's part, and it shows his instinct for seeing the difference between humans on the one hand and goods and machines on the other in experimental settings or modelling situations. It also points towards his later view which regarded program development as a social activity [19].

Nygaard later said [20] that "*my ambition was to build up OR as an experimental and theoretical science in Norway. I wanted our group to be reckoned as being among the top groups in the world in three to five years, and selected jobs and job strategies accordingly*". Nygaard was central in founding the Norwegian OR Society (NORS) in 1959, and chaired it the first five years. In 1960, acknowledging the achievements of Nygaard and his group, The International Federation of Operational Research (IFORS) asked NORS to arrange the 3rd international OR conference in Oslo.

Between 1959 and 1963 Nygaard's professional life changed completely. He was asked by representatives of Norwegian industry to establish a group for civilian OR at The Norwegian Computing Center (NCC) [16].[3] When Nygaard accepted the position at NCC in 1960, much to the chagrin of the NDRE Director, most of his NDRE OR group followed him[4]. In the autumn of 1963 Ole-Johan Dahl also joined Nygaard at NCC.

In May 1962 UNIVAC invited about 100 European computer people to the US to see their new UNIVAC 1107 computer. Nygaard accepted the invitation and took along a paper he had written about his early SIMULA ideas. The ideas caught on in UNIVAC and resulted in an 1107 offer to NCC at half price in return for a SIMULA compiler and a Linear Programming package that NCC planned. During some hard working summer months Nygaard, by establishing clever alliances, managed to convince the Royal Norwegian Council for Scientific and Industrial Research (NTNF) that the 1107 offer was a rare occasion of national importance, secured initial funding for the 1107, and in October 1962 had the UNIVAC and SIMULA contracts signed. The Oslo IFORS conference went successfully in June 1963 and the UNIVAC 1107 arrived in August.

From right: Dahl and Nygaard explain class and object to Sigurd Kubosch and Bjørn Myhrhaug, central implementors of UNIVAC and IBM mainframe SIMULA 67 compilers

This very intense period has been vividly described elsewhere by Bemer [1], Dahl and Nygaard [3], and Holmevik [9]. Holmevik also describes the cooperative process

---

[3] NCC was established in 1952 by NTNF as a national coordinator of computing for pure and applied research and re-established in 1958 as a research institute to promote the use of computers and quantitative methods.

[4] Nygaard had for some time been in conflict with the NDRE director who, according to Nygaard, didn't want an OR group to do scientific research and provide support for decision makers, but preferred researchers to point out "*the correct solution*" for the military establishment. He was "*happy to see a development that gave more power to his institute*" [20].

leading Dahl and Nygaard to the two versions of the SIMULA language, the ensuing work on compilers, and the spreading of the SIMULA ideas and its use. As for the complementing qualities of the two pioneers, Stein Krogdahl [11] who worked with them both, writes "*Nygaard had the idea to start with, and even if Dahl indeed participated in developing the concepts, Nygaard probably remained the one that most naturally expressed the high-level dreams and hopes for the project. On the other hand, Dahl was the one who immediately saw the implementational consequences of the different proposals, and could thereby mould them into a form that, to the largest possible extent, took care of both implementational and conceptual aspects*". Nygaard saw SIMULA as a language that would give him the concepts he needed for expressing system properties and describe his understanding to others, including the possibility to "*educate the machine*", i.e. run the description on a computer. For Nygaard, compilers and program execution were important and necessary, but after a while object-orientation for him was not about programming, it was a tool for modelling and understanding. That is what he devoted his later language work to after the initial success of SIMULA.

4 Working with Trade Unions

In the late sixties the Trade Union School (LO-skolen) asked Nygaard to lecture on modern technology. It was considered important that the workers should acquire the knowledge and competence necessary to influence the design of the workplace and their own working conditions. Nygaard was concerned that operations research and other IT-tools should not only be used for the benefit of the owners and the employers. The Norwegian Iron- and Metal Workers Union (NJMF) was at that time approached by many anxious local unions in connection with the introduction of numerically controlled machines.

In 1970 the NJMF union congress made the following resolution:

"*In connection with the development and use of computers, the congress underlines that a deliberate effort should be put forward to counteract the tendency to establish systems where humans are treated as a mechanical and programmable production factor. Where management is not willing to cooperate, the union must by themselves carry out such studies as will be necessary to strengthen the work place under the objectives of the workers and demand that the proposals that are brought forth shall be considered by the business management.*"

The time was therefore ripe when Nygaard suggested that NJMF should apply for research funding from NTNF [5]. The purpose of the application was "*to build a base of knowledge at least equal to that which is available to the employers*". This was the first time that a trade union had applied for funding for this kind of a project[5].

---

[5] The application was to be handled by the NTNF Research Committee for Mechanical Industry. The committee held its meetings in the premises of The National Mechanical Industry Organization (MVL), i.e. the employer organization for this sector of Norwegian industry. MVL did not want a project where NJMF cooperated with "*the radical researchers of the Norwegian Computing Center*". The NCC reputation was a result of the well published dispute between NCC and NTNF in 1969-1970 [4]. The leadership of NJMF, with the full support of the NJMF leader Leif Skau, opposed this attempt by MVL to move the project.

The project started at NCC in January of 1971 with Nygaard as the project leader [18]. Four local union shops, geographically spread over the southern part of Norway, were selected to take part. Eventually 120 people participated in the project.

It was important for NJMF and NCC that the project gave real and useful results. The Steering Committee in the autumn of 1971 formulated a new definition of project results: *"Results are all actions from NJMF, centrally or at the local level, that with support from the project aim to give the organization and its members a larger influence on data processing and control tools in their work place. ... From this viewpoint, working papers and research reports from the project will be useful only to the extent that they lead to actions of the organization's leadership, divisions, or local shops."*

The project group initially concentrated on establishing suitable educational material for shop stewards and industry employees. This resulted in a basic textbook for trade unions on "Data processing, planning and control" [17] which formed the basis for the work started by the shops of the participating companies from the summer of 1972. The purpose was that the shops should test a way of working that the project group thought would be common in the future, that shops themselves would study important questions within their work place with assistance from experts from outside and personnel from the company itself.

The Iron and Metal Project was a very untraditional research project. The contractor was of a new kind, the project had an uncommon definition of goals, and the group of participants was unusual. Both the establishment and the fulfilment of the project were wholly dependent on Nygaard's creative engagement. The project also demonstrated his ability to inspire co-workers and participants and to build a team. His lack of ability to follow the time-schedules given for project reports was also demonstrated[6].

The project was followed by similar projects and had important consequences for trade union involvement in the introduction and use of new technology in Norway, Scandinavia, and elsewhere [7].

Nygaard's last project initiative in cooperation with trade unions was UNITE (Union Net for Information, Teamwork and Education, 1983-85). The strategic importance of ICT for organizational development was the main theme. The focus was to study how the *"PC-revolution"* and data communication could make union work, particularly on the local level, more efficient and influential. As was often the case with Nygaard's initiatives, this project also had an international perspective and network building aspect.

5 Data Agreements and the Working Environment Act

In 1974 the world's first agreement on enterprise data use was established between the company Viking-Askim AS and the local shop [5]. The background was that the enterprise wanted to introduce a new data system for control of their car tire production. The trade union asked Nygaard for help in the discussions with the employer. The result was a data agreement that gave the employees a right to information and

---

[6] If Drude Berntsen, who was NCC's director at the time, did not have meetings with him twice a week to follow project reporting, the project would not have finished on time and hardly experienced such a successful end result.

participation in the design, introduction and use of data systems in the enterprise. Less than a year later The Norwegian Federation of Trade Unions (LO) and the Norwegian Employers' Federation (NAF) agreed on a "General Agreement on Technological Development and Computer Based Systems". A similar agreement was also established between the Government as well as municipalities and trade unions in the public sector.

In 1977 a new Working Environment Act was adopted in Norway. In §12 on The Organization of Work a new provision was introduced in the section Concerning Systems for Control and Planning: *"Workers and their elected representatives shall be kept continuously informed of systems used in planning and that are necessary to accomplish work, including planned changes to such systems. They shall be given the training necessary to familiarize themselves with the systems, and they shall take part in designing them."*

It is worth noting that for the most part, the introduction and use of data systems and technology in Norway went without conflicts and problems between employers and employees. This was largely a result of the early involvement in data policy by the trade unions, supported by Nygaard, and the ensuing introduction of §12 in the Working Environment Act.

6 Career as University Professor

After the Iron and Metal project Nygaard wanted to make what had been understood about the system development process and the societal implications of information technology, a part of academic teaching and research in information systems. For a year in 1974 he was guest professor at the University of Aarhus in Denmark and later on he was part time professor at the University of Oslo. There he worked in teams with students trying to build up an alternative curriculum in system development. This work is often referred to as the "Scandinavian School of System Development" and is closely linked to the field of Participatory Design.

Research in system development became also a major part of his work in the seventies. At NCC he played an important part in defining the DELTA system description language [8]. DELTA was a description tool, not a programming language. It became a useful platform for description of many aspects of the interactions between human actors and computing equipment. It was never widely used, but the ideas were carried on into new language activities.

Nygaard wanted to stay active both in traditional informatics and in system development [19]. By engaging in basic research he wanted to *"prove"* that he still was active in the scientific research community. While at Aarhus he therefore started a new basic research project in object-oriented language design. The language was called BETA [12]. BETA is built upon a few general, but very powerful concepts. By the mid eighties BETA was implemented on a series of computers.

He emphasised that object-orientation should be available in a system description and programming language because of its capability for system modelling and linking with the systems' environment, and also for its relevance to knowledge representation.

In 1984 Nygaard became a full time professor at the University of Oslo and in 1987 he was visiting professor at Stanford University, Palo Alto, USA, visiting scientist at Xerox Park in Palo Alto and a consultant and lecturer at Apple's Advanced Technology Group.

In the fall of 1988 "The Second European Conference on Object-Oriented Programming, ECOOP-88" was held in Oslo. It gathered close to 400 scientists from all over the world and was recognition to the fact that the cradle of object-oriented programming stood in Norway.

The project teams that Nygaard engaged in the development of BETA in Norway and in Denmark were later involved in creating ODSL [14] for the International Telecommunications Union and later in creating the de facto standard modelling language UML [26] that is used throughout the world today.

In the eighties Nygaard also became the chairman of the Steering Committee for a Cost-13 (EEC) financed European research project on the study of the extensions of profession oriented languages necessary when artificial intelligence and information technology are becoming part of professional work[7]. With this initiative he once again wanted to be in the forefront of the development.

After he retired from the University of Oslo in 1996, he became the leader of GOODS (General Object-Oriented Distributed Systems) [21], a 3-year Research Council of Norway (RCN) supported project, aimed at enriching object-oriented languages and system development methods with new basic concepts that make it possible to describe the relation between layered and/or distributed programs and the hardware and people carrying out these programs.

In 2002, shortly before Nygaard passed away, his last proposal received reluctant financial support from RCN. The COOL project (Comprehensive Object-Oriented Learning) was a 3-year interdisciplinary research project proposal launched by a consortium of four Norwegian research institutions, supported by research institutions in Aarhus, Denmark, with the intention of cooperating with test sites available through Nygaard's large personal international contact network. His idea was to produce an object-oriented "*Learning Landscape*" of pedagogical and organisational components. The proposal shows again the wide range in his ideas and that he always had visions for his work. After Nygaard passed away, the project was partly redefined and its global visions were toned down. Still, his inspiration and perspectives for COOL remained and resulted in a book offering the learner's perspective into the challenges of learning object-orientation [6].

7   Engagement in Politics

Nygaard had always been engaged in politics. He was active as a student and took part in strategic work for Venstre, a social liberal party at the centre of Norwegian party politics. He left the party at the end of the sixties because "*I started doubting my engagement in traditional party politics, and left the Liberal Party when I realized I had become a socialist*" [20]. In the sixties he worked and had positions in Naturvernforbundet, Norway's largest environmental conservation organization, and was

---

[7] A profession oriented language (POL) is a high-level language that, for the purpose of e.g. participatory design or system description and development, combines natural language concepts oriented towards a particular profession with concepts related to information processing [24].

engaged in work to establish alternative institutions for treating alcoholics and the homeless. During the intense political fight before the 1972 Referendum on whether Norway should become member of the European Common Market, he worked as coordinator for the large majority of youth organizations that worked against membership.

Nygaard joined the Labour Party in 1971; this membership lasted for about thirty years. In 2001 he left the party, *"disappointed by the right turn of the party"*. Nygaard was active in several Labour Party subcommittees discussing research policy and was a member of the party subcommittee on Data Policy (1980) [5].

In late 1988 he engaged himself as the chair of the Information Committee on Norway and the EEC, an organisation that was reorganized in August 1990 as "Nei til EU" (No to European Union Membership for Norway, NTEU). Nygaard was an extremely capable leader of NTEU. He managed to keep together an organization of people that belonged politically from the far left to the far right, people with highly different inducements for their opposition to EU membership. A prerequisite for such an organization to be trustworthy and have an impact was that it had a platform (a set of cornerstones formulated by Nygaard) that underlined its democratic ideological basis, as well as international orientation, and sharply disassociated itself from any racist attitudes.

Nygaard at a public meeting in Trondheim 1994, advocating democracy while opposing a union.

In a lengthy lecture first presented in Munich in 1995, Nygaard tried to explain his and NTEU's views on EU and EU membership to foreigners [25].

Before the Referendum Nygaard took part in a long line of public meetings and discussions with opponents and supporters of EU membership. He was a clever and knowledgeable debater and a worthy opponent to his foremost antagonist, Prime Minister Gro Harlem Brundtland. NTEU disseminated information from a critical point of view about Norway's relation to the Common Market and coordinated the efforts to keep Norway outside. Before the Referendum on November 28, 1994, NTEU had 145 000 members and succeeded in getting 52.2% of the votes.

8   From Conflicts to Recognition

Nygaard's views on conflicts and research were expressed in [20]: *"Has anyone resented the content of your work recently? If not, what is your excuse?"*

During his career Nygaard met several conflicts with what he termed as the research bureaucracy and *"the research-industrial power elite"*. He writes about his lack of popularity in this system and his many turned down applications for research funding [20]. He considered himself the Norwegian record holder in rejections.

The central people in NTNF were very sceptical about the further development of SIMULA in 1965-69. Without the profit from the sale of the surplus computer capacity of the UNIVAC 1107, SIMULA as an object-oriented language would never have taken place. Though Nygaard did not seek direct conflicts with funding authorities, he did not always show his diplomatic side when in their company[8].

His views on the relation between research, science, and politics were controversial, in particular in conservative circles. One episode recorded in [20] illustrates his opinions on this question[9].

However, in the years 1990-2002 Nygaard's work got recognition among his peers and in society that amply compensated for his many years of scientific and political seminal uphill activities. The list of prizes awarded him is impressively long, but what strikes one most is the spread of the awarding institutions. It started with the Norbert Wiener prize (1990) from The American Association of Computer Professionals for Social Responsibility, as the first non-US citizen, *"for his pioneering work in Norway to develop Participatory Design, which seeks the direct involvement of workers in the development of the computer-based tools they use."* He was appointed Doctor Honoris Causa at Lund University, Sweden (1990) and at Aalborg University, Denmark (1991), and received Computerworld Honorary Prize for *"making Norway internationally well-known in information technology"* in 1992. The Norwegian Data Association awarded Nygaard and Dahl its first Rosing Honorary Prize in 1999 and the Object Management Group awarded him an Honorary Fellowship for *"his originating of object technology concepts"* in 2000. Later that year, Nygaard and Dahl were both made Commander of the Order of Saint Olav by the King of Norway. There is no Nobel Prize for computer scientists, but when Nygaard and Dahl together received the ACM Turing Award in 2001 and the IEEE John von Neumann Medal in 2002, it was in recognition of work that clearly was of Nobel laureate stature. Proof of the wide international respect for Nygaard and his work is also demonstrated by the Nygaard memorial page kept by his institute [13].

After the untimely death of both Ole-Johan Dahl and Kristen Nygaard in 2002, Association Internationale pour les Technologies Objets (AITO) established an AITO Dahl/Nygaard Prize in their name. The prize is awarded annually for *"significant technical contributions to the field of Object-Orientation"*.

---

[8] In 1968, when NCC celebrated its tenth year as a research institute, Nygaard's principal speech reminded the notabilities present from NTNF about the relation between researchers and research funders: *"It is your money, but it is our lives."* This reminder was applauded by NCC employees, but did not go well with the high guests.

[9] *"I remember a lecture about the Iron and Metal Project, around 1974, for a group of very promising and very career-oriented executives in their mid-thirties. The atmosphere was reeking of hostility, and I got the question:*
   'Does not what you have done belong in politics rather than science?'
   'This question may be answered with Yes or with No,' I said. 'If you regard what you have learnt at the Norwegian Institute of Technology in Trondheim and the Norwegian School of Business Administration as belonging to politics, then what I have told you also belong to politics, and the answer is Yes.
   If you do regard what you have learnt there as science and not politics, then what we have done also belong to science, and the answer is No. Please, pick the answer you want'.
   I must admit that the answer was not appreciated."

9 Conclusion

Kristen Nygaard liked to be at the centre and to be seen, to show his qualities, and for them to be honoured. He was proud of his many honorary prizes. But he was also proud of his achievements, satisfied with their recognition, and he happily shared the recognition with his colleagues and co-workers. The hospitality of Kristen and his wife Johanna is widely acknowledged by all who came in contact with them.

The satisfaction gave him energy and desire to go on with his tireless efforts to spread the message that object-orientation is not just a technology for programming, but primarily a tool for modelling and organizing multi-perspective understandings of a system, thus contributing to further understanding. No task seemed too big for him, even at the age of 75.

The major chords of Nygaard's life's work are solidarity and empowerment. In his work on computer languages, system development, and object-oriented education, as well as in his work for the homeless, the trade unions, the environment, and the "Nei til EU" movement, he took the side of the common people against the forces of the market. He recognized that the means a person uses to structure his thinking about a phenomenon and his understanding of the world is not neutral. Information technology, he maintained, reflected by and

Kristen Nygaard and Ole-Johan Dahl in 2000, proud receivers of recognition from the King of Norway

Photo by: Siw Lene Ringvold

large the worldview of the market in terms of values, power and objectives. His work encompasses a critical re-examination of this, from the perspective of ordinary people, emphasising solidarity, democracy and decent working conditions.

References

In addition to the material referenced in the text and our private recollections, we have used Kristen Nygaard's home page [22] at the University of Oslo as a source for this presentation.

1. Bemer, B.: SIMULA, an ALGOL Offspring – First Object-Oriented Programming Language. In: Bob Bemer's Computer History Vignettes,
   http://www.bobbemer.com/SIMULA.HTM (last accessed January 12, 2010)
2. BIT (Nordisk Tidskrift for Informasjonsbehandling). Jan V. Garwick in Memoriam 29(3), 576 (1989)

3. Dahl, O.-J., Nygaard, K.: The Development of the Simula Languages. In: Wexelblat, R.L. (ed.) History of Programming Languages. Academic Press, New York (1981)
4. Elgsaas, K., Hegna, H.: The Norwegian Computing Center and the Univac 1107 (1963-1970). In: Bubenko, J., Impagliazzo, J., Sølvberg, A. (eds.) History of Nordic Computing: IFIP WG 9.7 First Working Conference on the History of Nordic Computing (HiNC1), Trondheim, Norway, June 16-18, 2003. Springer, New York (2005)
5. Elgsaas, K., Hegna, H.: The Development of Computer Policies in Government, Political Parties, and Trade Unions in Norway (1961-1983). In: Impagliazzo, J., Järvi, T., Paju, P. (eds.) History of Nordic Computing 2: IFIP WG 9.7 Second Working Conference on the History of Nordic Computing (HiNC2), Turku, Finland, August 21-23, 2007, Springer, Boston (2009)
6. Fjuk, A., Karahasanovic, A., Kaasbøll, J. (eds.): Comprehensive object-oriented Learning.: the Learner's Perspective (COOL). Informing Science (2006), ISBN 8-392-23374-3
7. Floyd, C., et al.: Out of Scandinavia: Alternative Approaches to Software Design and System Development. Human-Computer Interaction 4(4), 253–350 (1989)
8. Holbæk-Hanssen, E., Håndlykken, P., Nygaard, K.: System Description and the DELTA Language. In: DELTA Project Rep. No. 4, NCC Publ. No. 523. Norwegian Computing Center, Oslo (1975)
9. Holmevik, J.R.: Compiling Simula: A Historical Study of Technological Genesis. IEEE Annals of History of Computing 16(4), 25–37 (1994)
10. Kay, A.C.: Interview in The Norwegian Broadcasting System (NRK) P2, Oslo, Norway, December 26 (2002)
11. Krogdahl, S.: The Birth of Simula. In: Bubenko, J., Impagliazzo, J., Sølvberg, A. (eds.) History of Nordic Computing: IFIP WG 9.7 First Working Conference on the History of Nordic Computing (HiNC1), Trondheim, Norway, June 16-18, 2003. Springer, New York (2005)
12. Lehrmann Madsen, O., Møller-Pedersen, B., Nygaard, K.: Object-Oriented Programming in the BETA Programming Language. Addison-Wesley/ACM Press (1993), ISBN 0-201-62430-3
13. Memorial Site for Kristen Nygaard, http://www.ifi.uio.no/in_memoriam_kristen/ (last accessed January 21, 2010)
14. Møller-Pedersen, B., Haugen, Ø., Belina, F.: Object-Oriented SDL. Tele (Swedish Televerket's technical journal) (January 1991)
15. Norsk matematisk tidsskrift, vol. 27 (1). Resultatet av premiekonkurransen for gymnasiaster, Oslo (1945)
16. Holden, L., Hegna, H. (eds.): Norsk Regnesentral 1952–2002 (The History of the Norwegian Computing Center 1952-2002), Norsk Regnesentral, Oslo (September 2002) (in Norwegian)
17. Nygaard, K., Bergo, O.T.: Planlegging, styring og databehandling. In: Grunnbok for fagbevegelsen, Planning, Control and Data Processing. Basic Reader for Trade Unions, 2nd edn. Tiden Norsk Forlag, Oslo (1973)
18. Nygaard, K.: The Iron and Metal Project. Trade Union Participation. In: Proceedings of the CREST Conference on Management Information Systems 1977. Cambridge University Press, London (1977)
19. Nygaard, K.: Program development as a social activity. In: Kugler, H.-J. (ed.) Proceedings from the IFIP 10th World Computer Congress, IFIP, Dublin, Ireland, September 1-5. Elsevier Science Publishers B.V, North Holland (1986)

20. Nygaard, K.: Those were the days. Or "Heroic times are here again? In: Opening lecture at The Information Systems Research in Scandinavia (IRIS) Conference (August 1996); Scandinavian Journal of Information Systems 8(2), 91-108 (1996)
21. Nygaard, K.: GOODS to Appear on the Stage. In: Aksit, M., Matsuoka, S. (eds.) ECOOP 1997. LNCS, vol. 1241, pp. 1–31. Springer, Heidelberg (1997)
22. Nygaard, K.: Home page, http://heim.ifi.uio.no/~kristen/ (last accessed January 21, 2010)
23. Nygaard, M.: Notes on Kristen Nygaard's early years and his political work. In: Böszörmenyi, L., Podlipnig, S. (eds.) People behind Informatics, in memory of Ole-Johan Dahl, Edsger W. Dijkstra, and Kristen Nygaard, Eigenverlag Universität, Klagenfurt, Germany (August 2003); Re-printed for 18th European Conference on Object-Oriented Programming, Oslo, Norway (2004)
24. Nygaard, K.: Profession Oriented Languages In: Bøgh Andersen, P., Bratteteig, T. (eds.) SYDPOL Program working group 2: Computers and language at work: the relevance of language and language use in development of computer systems. Research Report Series, vol. 126, Institute of Computer Science. University of Oslo, Norway (1989)
25. Nygaard, K.: We are not against Europe, we are against Norwegian membership in the European Union. Lecture at "Europa Fluch oder Segen?" Münich (January 1995), http://heim.ifi.uio.no/~kristen/POLITIKKDOK_MAPPE/P_EU_Munchen_eng.html (last accessed May 20, 2010)
26. UML, Unified Modelling Language, http://www.uml.org/ (last accessed January 24, 2010)

Appendix: Abbreviations Used:

NDRE	The Norwegian Defence Research Establishment
IFORS	The International Federation of Operational Research
NORS	The Norwegian Operational Research Society
NTNF	The Royal Norwegian Council for Scientific and Industrial Research
RCN	The Research Council of Norway – RCN replaced NTNF and the other Norwegian research councils in 1993
NCC	The Norwegian Computing Centre
LO	The Norwegian Federation of Trade Unions
NAF	The Norwegian Employers' Federation
MVL	The National Mechanical Industry Organization
NTEU	No to European Union Membership for Norway
NJMF	The Norwegian Iron and Metal Workers Union
EEC, EU	European Common Market, European Union

Projects and Activities of the IPSJ Computer History Committee

Eiiti Wada

IIJ Innovation Institute,
Jinbocho Mitsui Bldg., 1-105, Kanda, Jinbo-cho, Chiyoda-ku, Tokyo, Japan
wada@u-tokyo.ac.jp

Abstract. IPSJ (Information Processing Society of Japan), which was launched in 1960, formed the Computer History Committee in 1970 in order to record the early history of Japanese computers. The Committee has continued its activities, publishing history related books, maintaining the Virtual Computer Museum web pages, editing articles about old computers and technologies, etc. Recently, it started another mission, searching and nominating the Information Processing Technology Heritages. The present paper surveys the Committee's long time loci in a concise way.

Keywords: early computers in Japan, Virtual Computer Museum web page, oral history, Information Processing Technology Heritage.

1 Introduction

"An early computer was very much a thing", scribed Professor Maurice Wilkes of Cambridge University at the beginning of his book [1]. Yes indeed! In fact, computers have been very much things until recently when people began carrying personal computers with no idea about the inside of their machines.

Early computers were real excitement for researchers, developers, even for users employing computers for their number crunching. In those good old days, the computers were physically very imposing. The arrays of vacuum tubes were registers directly in the engineers' sight. Mystically twinkling patterns on the cathode ray tubes incarnated the program loops. Irregular sounds of relay contacts helped locate program bugs. Architecture models of computers were implemented in exactly the same form. These are now merely folklore.

In order for young students to share the excitement of this dawn of computers, a number of activities are conducted on a worldwide scale. Here in Japan, Professor Ryota Suekane (1925-1987) started the Computer History Committee in IPSJ (Information Processing Society of Japan) in 1970 aiming to collect records of the development activities of early Japanese computers. The Committee stopped work once this original goal was fulfilled. Years later, Professor Shigeru Takahashi (1921-2005) reconvened the Computer History Committee, with the primary goal to edit "*The History of Japanese Computers*," which was published in 1985.

A. Tatnall (Ed.): HC 2010, IFIP AICT 325, pp. 50–57, 2010.

Since that time, the Computer History Committee continually maintained their activities in many dimensions, wishing to transfer the precise historical record of what took place in this country for newcomers in the information processing field. The purpose of this short article is to report on our 40 years of work.

2 The History of Japanese Computers Series

The first volume of the series covered the material up to 1960, when computer research emerged in universities, laboratories, as well as in a few industries. This volume included a number of parts. In one part, pioneering works were introduced; for instance, switching theory in mid 1930's by Akira Nakajima, who worked with NEC, the first generation computers like FUJIC (vacuum tubes, Fuji Photo Industry), parametron computer PC-1 [2], vacuum tube computer TAC (both at the University of Tokyo) and others are explained. Programming technologies of those days are the subjects of another part.

Following that, the second volume, *"The History of Japanese Computer Development"* was published in 1998 covering the era from 1960 to 1980. As a matter of course, it included many more commercial computers than the first volume. At the time of publication, the first volume was completely sold out. Accordingly, a compact disc containing the contents of the first volume was included in the second for the readers' convenience.

The Committee is now in the midst of editing of the third volume, to be published in 2010, commemorating the jubilee of the Society. The third volume covers facts of time span 1980 to 2000. Among newly added topics are, for example, word processors, computer networks and the Internet. This volume is likely to become much thicker than its predecessors.

3 The First Transactions

Much earlier than these books, on the occasion of the 10th anniversary, IPSJ and the first Computer History Committee planned to have a series of interviews of the computer pioneers, to be included in the national historical archives, in order that those precious achievements would not become vapour. The interviews came out in the form of 10 articles and appeared in the IPSJ Magazine from 1974 to 1978. Topics and authors with volumes and numbers are: FUJIC (B. Okazaki, **15**, 8), PC-1 (E. Goto, **16**, 1), MUSASINO-1 (K. Takashima, **16**, 2), ETL Mark III and IV (S. Takahashi, **17**, 2), ETL Mark I and II (Y. Komamiya, **17**, 6), NEC (H. Kaneda, **17**, 9), TAC (K. Murata, **18**, 3), Fujitsu (T. Matsuyama, **18**, 7), Oki (N. Sugiura, **19**, 5), and Hitachi (T. Uraki, **19**, 8).

4 The Virtual Computer Museum

In the early 1990's, Professor S. Takahashi suggested opening a Virtual Computer Museum on the Internet. The intent was to design new web pages showing many of the computers, peripherals, mini and personal computers, and so on, with photographs and precise descriptions.

In all, eleven categories for the Computer sections were provided. Those categories were: Early Computers, Mainframes, Supercomputers, Office Computers, Minicomputers, Workstations, Personal Computers, Japanese Word Processors, Other Computers, Peripheral Equipments, and Operating Systems. If we peruse the Early Computer web pages, there are the chronological lists, and early works by Ryoichi Yazu, who developed his mechanical calculator in 1902, by Torajiro Omoto, who started a manufacturing and marketing company for mechanical calculators as early as 1923, etc. Both Japanese and English pages were prepared. The web pages were continually kept updated and were enhanced by the addition of new items found since the initial release.

How many items are there in the Virtual Computer Museum? I counted the lines of the index and got an answer. About 950 items are included in the web site in all. (See http://museum.ipsj.or.jp/en/)

There are also other web pages for computer pioneers who endeavoured to realize the practical computer world. About sixty or so people were chosen from the indices of the above mentioned books of history as computer pioneers, and their prominent works in the computer field were uploaded to the web pages, both in Japanese and in English. If the pioneer was still alive, the text was written by the pioneer himself, otherwise, one of the disciples or colleagues was asked to contribute.

5 The Oral History Interviews

One urgent mission for the Committee is the oral history collection, i.e., the actual words spoken by each of the computer pioneers. In fact, less than half of the computer pioneers are still alive today. Sadly, some of those are already too aged to participate in interviews. The oral history subcommittee selects three to five candidates each year, makes arrangements for the interview, and listens to and records the story. Up to now, about 30 pioneers have already been interviewed.

As expected, the hardest part of this activity is in editing into the final form. Therefore, many records are still in the editing queue.

6 "My Poetry and Truth," Pioneers Reminiscences

The number of interviewers for the oral history project averages about five. Only limited members are allowed to join. On the other hand, the story told by a pioneer often tends to be quite interesting, and worth being listened to, and which, if possible, should be heard by a wider audience. From these observations, the Committee proposed to have special sessions at the Annual Congress of the Society. On each occasion, a couple of pioneers are given slots to lecture on the works of their younger days.

Following the autobiography by Johann Wolfgang von Goethe, *"From My Life: Poetry and Truth"* (in German, *"Aus meinem Leben: Dichtung und Wharheit"* '), the session was titled as *"My Poetry and Truth"*.

In 2008, the lecturers were Professors S. Noguchi and M. Nagao; in 2009, Professors K. Mori and K. Ikeda gave talks. We will have this session for a third time in March 2010 when ISPJ holds the Congress. The talks by Professor S. Mizutani and Honorable Chairman Mr. T. Yamamoto of Fujitsu Ltd. are now on the program.

7 The Information Processing Technology Heritages and Satellite Museums

Besides the Virtual Computer Museum, the Committee has long yearned to have a real museum, somewhat like the Computer History Museum in Mountain View, California.

Recent computers are too tiny to see the computing mechanism. On the contrary, ancient machines were very much instructive for studying architectures, memory elements etc. Therefore, those machines are quite precious for preserving information technology culture.

On the other hand, old machines have disappeared so fast. When those machines ceased operation, they were instantly turned to useless debris, and it is quite natural for the administrators of machines to throw them away to use the area for modern successors.

While planning to open the real computer museum, the Committee suddenly realized the need to collect the machines to display. Famous machines were disappearing from sight too quickly; and therefore throwing them out should be stopped as urgently and as soon as possible.

The conclusion by the Committee is to select valuable machines as Information Processing Technology Heritages, issuing certificates and ask the administrators not throw away these machines. The selection of heritage machines started in 2008. On March 2, 2009, the first set of heritage computers, or hardware parts, were nominated.

Soon after the nomination ceremony, the Committee restarted the selection of new heritage machines for the fiscal year 2009. The list of heritages and satellite museums will be announced in March 2010.

7.1 Information Processing Technology Heritages

The Information Processing Technology Heritages nominated for the year 2008 are as follows:

- Analogue Calculator for Simultaneous Equations of Order 9 (see the later section)
- ETL Mark II (relay computer at the Electro Technical Laboratory)
- ETL Mark IV Plug-in Packages and a Magnetic Drum
- FACOM128B (relay computer still in working condition, Fujitsu make)
- FUJIC (vacuum tube computer)
- H-8564 Magnetic Disk Drive
- HITAC 10 (one of the popular minicomputers)
- HITAC 5020 (mainframe developed by Hitachi)
- JW-10 (the first Japanese word processor, Toshiba)

- Jido Soroban (automatic abacus invented by Ryoichi Yazu in 1902. see the later section)
- Kawaguchi Style Electric Tabulation Machine and Turtle-shape Perforator (see the later section)
- MARS-1 (Japan National Railways seat reservation machine)
- MUSASINO-1B (parametron computer, Nippon Telegraph and Telephone)
- NEAC Series 2200 Model 50
- NEAC-2203
- OKITAC-4300C
- OKITYPER-2000
- Osaka University Vacuum Tube Computer (one of 3 vacuum tube computers)
- PC-9801 (personal computer by NEC)
- Parametron (logical component invented by E. Goto in 1954)
- SENAC-1 (NEAC-1102, parametron computer by NEC)
- TOSBAC-3400
- Tiger Calculator No.59 (mechanical calculator made in 1923. see the later section)

(See http://museum.ipsj.or.jp/en/heritage/index.html)

7.2 The Satellite Museums of the Historical Computers

Similarly, the Committee chose two institutions known to possess many historical computers as the satellite museums of the future, and also a real computer museum which will become the central office.

Satellite Museums are:

- KCG Computer Museum, Kyoto Computer Gakuin (institute)
- The Nishimura Computer Collection (personal collection of Professor H. Nishimura)

(See http://museum.ipsj.or.jp/en/satellite/index.html)

8   Articles and Series Appeared in the IPSJ Magazines

IPSJ publishes the Magazine for members of the Society monthly. From time to time, the editorial board plans to publish historical materials.

As mentioned, the first series were about early machines contributed by the members of the projects. Other special issues or series will be mentioned below.

8.1   Special Issue of "Less Known Computers"

In 1970's and 80's, many novel architectures were suggested, investigated and implemented such as Lisp machines, data-flow computers, and inference machines for the New Generation Computer Project (the Fifth Generation Computer). Because the number of first generation computers was small, and because they were seen as curiosities, there were the opportunities for them to be described fully, except for the commercial machines. Thus, the special issue was edited to publicize the idiosyncratic computers which would have been otherwise forgotten. The special issue *"Less Known Computers"* appeared in February 2002.

Only the names of these computers and authors are listed here: HITAC 2010 (S. Takahashi, T. Uraki), QA-1 (S. Tomita), Lisp machine and Prolog machine of Kobe University (Y. Kaneda et al), PACS (C. Hoshino), ELIS (Y. Hibino), EVLIS machine (H. Yasui), FLATS (T. Soma), FACOM α (H. Hayashi), SIGMA-1 (K. Hiraki), PIE (H. Tanaka), EM-4 (S. Sakai) and SM-1 (T. Yuasa).

The pdf's of those articles maybe obtained from the Virtual Museum web site.

8.2 "The Trail of Information Processing Technology in Japan" Series

In October 2002, a new series of the computer history related articles started. The series tried to include much wider topics. The summary of titles and authors are as follows:

- *Advances in Kanji/Japanese Processing Technologies: Input/Output Method in the Early Stage* (T. Uraki)
- *Advances in Kanji/Japanese Processing Technologies: Kana-Kanji Transfer Technology* (R. Kobayashi)
- *Advances in Kanji/Japanese Processing Technologies: The Birth and Brief History of the Japanese Wordprocessor* (S. Amano, K. Mori)
- *Advances in Kanji/Japanese Processing Technologies: Standardization of Coded Kanji Set* (K. Shibano)
- *Japanese Semiconductor Technologies for Computers* (Y. Tarui)
- *History of Design Automation in Japan* (A. Yamada)
- *Rise and Fall of Plug-Compatible Mainframe (1-3)* (S. Takahashi)
- *Advance of Computer R&D at NTT* (I. Toda, T. Matsunaga)
- *History and Deployment of the Information Technology Standardization Activity in the Information Processing Society of Japan* (A. Tojo)
- *The Seven Dwarfs and Japanese Computer Makers (1-2)* (S. Takahashi)
- *MITI (Ministry of International Trade and Industry) and Japanese Computer Makers* (S. Takahashi)
- *Topics on the Japanese Processing Technologies: The Stories of Japanese Processing, JEF and OASYS* (Y. Kanda)
- *Topics on the Japanese Processing Technologies: A History of Japanese Information Retrieval Technologies* (H. Fujisawa, H. Kinukawa)
- *Topics on the Japanese Processing Technologies: Development of Japanese Processing - BUNGOU, JIPS, M Method Input System* (H. Itoh)

These 17 articles were contributed and appeared on the Magazine from October 2002 until January 2004. The pdf's of those articles are placed on the Virtual Museum web pages.

8.3 Anatomy of Heritage Computers

Of twenty three nominated heritage machines and hardware parts, the older items are naturally analogue and/or mechanical. It is beyond our understanding to know how they really worked.

Taking this opportunity, the Magazine of the Society started a more or less short series of articles to analyze those unfamiliar computers. The first article was devoted to the analogue simultaneous linear equations solver, that is presently displayed at the National Science Museum.

This is the only surviving machine of the family. The original machine was constructed at MIT around 1936 by John B. Wilbur based on the ingenious idea of William Tomson far back in 1878.

In the next article, the pinwheel mechanism and related technologies of the Tiger Brand mechanical calculator developed in 1923 in Japan were described.

The third article discussed the Kawaguchi style tabulation machine, which is displayed at the Statistical Research and Training Institute, Statistics Bureau. In one sense, the machine is similar to the Hollerith tabulator. However, Ichitaro Kawaguchi installed various unique ideas to his machine. For instance, it has card bins to catch the falling cards from the sensor station placed on the top of the towery box.

The last article of the series will treat Yazu's mechanical calculator. Although the adder employs the standard Odhner type mechanism, the pin action of Yazu's machine is unexpected, moving left and right to engage and disengage with the wheels. In place of the multiplier register, it has the reverse rotating rod to decrease the multiplier from the product register.

The Committee hopes that, through these articles, the unique and interesting aspects of these historical machines may be exposed to the wider computer society.

The current series consists of four articles altogether, of which the first one appeared in September 2009. The last article will come out in March 2010.

9 Demos and Symposia

In this section, historical events which occurred outside of the Committee, but inside of the Society are summarized.

Taking the opportunity presented by the IPSJ Annual Congress, demonstrations of the old machines were held twice. The demo area was in a corner of the Congress and a small number of the old heroic machines or their components were displayed. Unfortunately, old computers were, needless to say, gigantic, and hard to transport. So sometimes only a picture panel was shown instead. They were quite enough to give the audience a feel of the deep impact and impression of these classic relics.

IPSJ operates a long sequence of Programming Symposia; in fact the first symposium was held in January 1960, shortly before the founding of IPSJ. These regular symposia are held in January with no restrictions on the topics. On the other hand, in summer, minor but intensive symposia are held on specific subjects.

In one special symposium, papers concentrated on computer history were solicited. The contributed papers can be seen on the Virtual Computer Museum web pages.

At its 40th symposium, Professor S. Okoma was invited as the guest speaker. Being one of the computer historians, he gave a talk on the current status of computer history research.

10 Conclusion

IPSJ has many committees under its umbrella. Most of them are research oriented. Members are relatively young. On the other hand, although there might be computer history research, the goal of the Computer History Committee, with senior constituents, is not in research but in keeping computer history alive. The Committee consumes most of its energy collecting and recording the long range of computer history. We edited a series of books on computer history, uploaded web pages for the Virtual Computer Museum, collected oral history interviews, planed special tracks at the Society's Congress for pioneering works, and published relevant articles in the IPSJ magazines.

Last, but absolutely not least, our central mission is to establish real computer museum in the near future, ever overcoming many hard obstacles. Although their realization is far from assured, every slight possibility has to be pursued.

The real computer history museum is not yet here. Nevertheless the large number of accesses to our Virtual Computer Museum web pages encourages the Committee, because so many people indicate their interest in the history of computers. The Computer History Committee makes up its mind anew to pursue the dreams for the real museum.

References

1. Wilkes, M.V.: Time-Sharing Computer Systems, Macdonald (1968)
2. Wada, E.: The Parametron Computer PC-1 and Its Initial Input Routine. In: Rojas, R., Hashagen, U. (eds.) The First Computers - History and Architectures. MIT Press, Cambridge (2000)

Contested Histories: De-mythologising the Early History of Modern British Computing

David Anderson

University of Portsmouth, "The Newmanry",
36-40 Middle Street, Portsmouth, Hants, United Kindom, PO5 4BT

Abstract. A challenge is presented to the usual account of the development of the Manchester Baby which focuses on the contribution made to the project by the topologist M.H.A. (Max) Newman and other members of the Dept. of Mathematics. Based on an extensive re-examination of the primary source material, it is suggested that a very much more significant role was played by mathematicians than is allowed for in the dominant discourse. It is shown that there was a single computer-building project at Manchester in the years immediately following World War II and that it was conceived, led, funded, supplied and staffed by Newman who was supported throughout by his long-time friend P.M.S. (Patrick) Blackett. In the course of the paper three persistent myths, which lend support to the dominant narrative, are identified and debunked.

Keywords: Manchester Baby, SSEM, Max Newman, Patrick Blackett, British Computing, Historiography.

1 Introduction

It is almost exactly 60 years since the world's first digital electronic stored-program computer – the Manchester Baby or SSEM -ran for the first time. The story of how the machine came to be developed by Prof F.C. Williams and Mr. T. Kilburn from the Dept. of Electro-Technics, without any significant external assistance is well known and has come, within the sub-discipline of the history of computing, to represent a dominant historical narrative.

I will offer a challenge to the usual account of the development of the Manchester Baby. In doing so, I will focus on the contribution made to the project by the topologist M.H.A. (Max) Newman and other members of the Dept. of Mathematics. Based on an extensive reexamination of the primary source material, I will suggest that a very much more significant role was played by mathematicians than is allowed for in the dominant discourse. In short, I will show that there was a single computer-building project at Manchester in the years immediately following World War II and that it was conceived, led, funded, supplied and staffed by Newman who was supported throughout by his long-time friend P.M.S. (Patrick) Blackett. I will identify and debunk three persistent myths, which lend support to the dominant narrative.

A. Tatnall (Ed.): HC 2010, IFIP AICT 325, pp. 58–67, 2010.

2   A Note on Motivation

It is perhaps worth saying explicitly that my purpose is not to argue that historians of computing should, in their accounts of the period, replace heroic engineering pioneers with heroic mathematicians. The principal problem with the conventional account is not that it allocates credit inappropriately but that in so doing it helps obscure much more interesting historical questions concerning the role played by the government in fostering and directing technological innovation and development in Britain both during the second world war and in the immediate post-war period.1 The historical re-evaluation of the development of the Manchester Baby which I present should not be understood as an end in itself nor, primarily, as a contribution to a credit dispute but rather as a necessary clearing of the ground so that historians of technology can engage with topics of greater historical significance.

3   M.H.A. Newman

Despite having received very little attention from historians of computing, it is no exaggeration to say that Max Newman was one of the most significant figures in the early history of British computing. His direct influence was exerted over more than a decade beginning at Cambridge before the Second World War, continuing at Bletchley Park during hostilities, and finishing in the peace-time setting of Manchester in the mid-late 1940s.

Newman's deeply-ingrained habit of understating his own contribution and preference for stressing the accomplishments of others goes some way towards explaining why historians of computing have generally paid only superficial attention to this remarkable man who is, in consequence, principally remembered today for his work as a mathematician and topologist.

4   Turing and the Roots of Modern Computing

Modern computing is often said to have originated with A.M. (Alan) Turing. If so then its roots can be traced back through Newman to a talk given by David Hilbert at the Sorbonne on the morning of the 8th August 1900 in which he proposed twenty-three "future problems" for mathematics research in the 20th century. The tenth of Hilbert's questions led directly to Hilbert and Ackerman's 1928 formulation of the Entscheidungsproblem [1], which Hilbert considered to be "the central problem of mathematical logic" [2]. The essence of the question was: could there exist, at least in principle, a definite method or process involving a finite number of steps, by which the validity of any given first-order logic statement might be decided?

Turing seems first to have encountered the Entscheidungsproblem around the Spring of 1935 when he was a student on Newman's Part III course on the foundations of mathematics [3]. Solving the Entscheidungsproblem rigorously was entirely dependent on the extent to which a formalisation of the notion of "process" could be devised and it was this task which Turing, acting on a suggestion of Newman's, accomplished so dramatically:

In the middle of April 1936, Turing presented Newman with a draft of his breath-takingly original answer to the Entscheidungsproblem [4]. At the heart of Turing's paper was an idealised description of a person carrying out numerical computation which, following Church, we have come to call a "Turing machine". All modern computers are instantiations of Turing machines in consequence of which Turing's paper is often claimed to be the single most important in the history of computing.

From the moment Newman saw Turing's solution he took him under his wing. Newman canvassed successfully for "On Computable Numbers" to be published by the London Mathematical Society and, simultaneously, enlisting Alonzo Church's assistance in arranging for Turing to spend some time studying in Princeton.

Cambridge in the late thirties and early forties seems to have provided particularly fertile soil for computing pioneers and Newman played a part in the education of most of them. In addition to Alan Turing and his exact contemporary Maurice Wilkes, other students of Newman's included Tom Kilburn, Geoff Tootill and David Rees.

5 Bletchley Park

On the 16th March 1939, as war was breaking out across Europe, Newman was awarded a fellowship of the Royal Society. However, there was little opportunity to use this as a springboard for further work in mathematics. The outbreak of hostilities took one colleague after another out of academic life into war work. Newman grew increasingly disillusioned with life at Cambridge and at the suggestion of Blackett, he accepted a post at Bletchley Park. Neither of them could have had the least inkling that Newman had embarked on a course which was to completely alter the future direction of his career. Max was initially appointed as a cryptanalyst as part of John Tiltman's group. The type of transmission which attracted the greatest interest was known as 'Tunny' and carried messages between the very highest ranks of the German command. Manual methods utilising statistical techniques had been devised for breaking into the code but the sheer volume of traffic being intercepted was beginning to overwhelm the human resources available.

Newman believed it was possible, in principle, to mechanise the attack on Tunny and successfully lobbied to test his conviction by developing an electro-mechanical code-breaking machine which came to be known as the "Heath Robinson".

The Heath Robinson proved fairly unreliable but the results it achieved were sufficiently impressive for approval to be given to develop a more sophisticated machine – the Colossus. A great deal has been written about this apparatus now widely recognised as the world's first digital electronic computing machine. In the current context, it should be sufficient to note that had Newman done nothing else in his career he would have been assured of a place among the most important figures in early British computing simply by virtue of having led the development of this machine.

6 Manchester and the Birth of Modern Computing

The Colossus was to have a profound effect on Newman's future career. He saw at once, as few others did, the impact that computing would come to have on mathematics and resolved to establish a computer-building project as soon as the war was over. The mathematics department at Cambridge was not the right environment in which Newman's new ambition could be pursued and he began to look around for a more favourable setting. With the help and encouragement of Blackett, Max was appointed to the Fielden Chair of Pure Mathematics at Manchester University. Newman had two clear goals in mind: to establish a first rate department which could stand comparison with the best in the country and to build a computer. At Bletchley, Newman was surrounded by people who could help him achieve both objectives.

In a clear declaration of intent, Newman brought with him from Bletchley Park, I.J. (Jack) Good and David Rees11. Both these Cambridge mathematicians had served in the Newmanry and in addition to having impeccable mathematical qualifications both had a familiarity with the Colossus that would be invaluable for the work that Newman had in mind. With Blackett's assistance a substantial grant was secured from the Royal Society explicitly for the purpose of developing a computer – the first such award ever made and a huge triumph for Manchester.

The only piece of the puzzle that was missing was a lead engineer. Newman had, before submitting his funding application to the Royal Society, secured limited support from Prof Willis Jackson who was the head of Manchester's department of Electro-Technics. Jackson agreed that when Newman was able to secure the services of a full-time engineer he could be attached to Jackson's department. It is apparent from the available documentary evidence that Newman felt able to take Jackson into his confidence to a much greater extent than is generally supposed. The work carried out at Bletchley Park during the Second World War was classified well beyond Top Secret and Newman was very well aware of the restrictions imposed on him by the Official Secrets Act. The very great extent to which the secrets of Bletchley Park were preserved has often been remarked upon, indeed Churchill described the Bletchley code breakers as his "geese that laid the golden eggs and never cackled". Within the History of Computing it is usually claimed that the Colossus could have had no impact on the development of civilian computing because even the mere fact of its existence was kept completely secret until the mid-1970s. I have elsewhere called this claim the myth of secrecy [5].

In fact, Newman not only revealed the existence of Colossus to Willis Jackson but actually took him, during the Summer of 1945, to see a number of Colossi in situ at Bletchley Park and obtained advice from him on which components of the machines could be re-used in the construction of a computer at Manchester. Acting directly on Jackson's advice Newman made arrangements for "...the material of two complete Colossi"[6] to be transported from Bletchley to Manchester – the transfer taking place later the same year. The significance of Jackson's visit to Bletchley is enormous since it could not have taken place without security clearance having been obtained at the highest level. Jackson's presence in Bletchley during 1945 is a clear indication of government support for Newman's computer-building project and constitutes conclusive disproof of the claim that Colossus could not have had an impact on the development of peacetime computing because of the secrecy surrounding its existence. In fact there is also some documentary evidence supporting the claim that Newman discussed the Colossus with John von Neumann and with F.C. Williams. There are also clues in the Royal Society papers concerning Newman's funding application that his involvement with computing during the war was known to the funding committee.

Jackson's support for Newman's project was valuable but stopped short of providing him with the engineering expertise he required to actually construct a computer. Newman was not the only person looking for a top-flight engineer; the National Physical Laboratory(NPL), were also planning to build a computer and "Good circuit" men, as Newman wrote to von Neumann, were "both rare and not procurable when found" [7]

F.C. (Freddie) Williams found himself at the end of the war in the fortunate position of being a man greatly in demand. Williams had been a lecturer at Manchester before the war and had spent the war years working at the Telecommunications Research Establishment helping develop Radar and leading a small trouble-shooting team. By the end of hostilities Williams was widely recognised as one of the best electrical engineers around and he was actively courted by NPL and by Newman. In the end Williams choice was relatively easy to make. NPL's proposal was that Williams should work on Turing's ACE design and not only afforded Williams less freedom to develop his own ideas than he wanted but also compelled him to work with Turing -a prospect Williams did not relish. Newman and Manchester were able to offer Williams the prospect of a chair in Electro-Technics On his appointment, Williams brought with him, Tom Kilburn, newly re-cast as a Ph.D. student funded by TRE. Having secured the support of the university, obtained funding from the Royal Society and assembled a first-rate team of mathematicians and engineers, all the elements of Newman's computer-building plan were in place. Adopting the same approach as he had used at Bletchley Park, Newman set his people loose on the detailed work while he concentrated on orchestrating the endeavour. The result was, once again, success beyond all expectation. By the middle of 1948 the Manchester Baby was up and running and although it was little more than a proof of concept it was still the world's first working digital electronic stored program computer. Through the agency of Patrick Blackett, Sir Ben Lockspeiser was shown the Baby and government support for the manufacture by Ferranti of a production version of the machine was quickly secured.

7 Contested History

The account which I have given of the development of the Manchester Baby and the role played in it by Newman represents a very radical departure from the conventional history of the project. It is appropriate now to consider in some detail the principal ways in which the dominant narrative differs from the interpretation which I have presented.

The earliest history of the development of the Baby was written by S.H. (Simon) Lavington according to whom there were, in fact, two distinct and separate Manchester projects to build a stored program digital computer, one led by Williams and the other by Newman. On this account Williams' project is presented as a triumph from which the engineers emerge as heroes whereas, by contrast, Newman's project is characterized as a failure and the mathematicians, in so far as they are mentioned at all, are portrayed as marginal figures.

This "two project" account has very general currency and forms part of a wider professional mythology within which the engineering or practical perspective is privileged while the mathematical or theoretical point of view is almost entirely excluded. Lavington's assertion that the mathematicians took no active part in the design process sounds plausible. However the claim of non-involvement is more complex than may, at first, be apparent and is entirely dependent on what precisely is meant by 'active' and 'design'. Williams' personal recollection after the event that neither he nor Kilburn knew the first thing about computers until 1947 when Newman and Turing explained to them how computers work [8] is prima facie evidence, coming from an engineering source, of at least one active contribution made by the mathematicians. Further evidence of collaboration is to be found in the contemporaneous records kept by Jack Good, one of Newman's mathematicians, which show a free exchange of ideas and documents passing between the engineers and the mathematicians. Good also recalled having had a hand in the general theoretical education of the engineers. According to Donald Michie, Newman delivered, at Manchester in the immediate post war period, a series of lectures on computing which helped shape the understanding of the engineers and constituted part of their computing education [9].

Moving outside the immediate Manchester circle, Turing and Wilkinson delivered a series of lectures on the design of the ACE which took place at the Ministry of Supply's London Headquarters from late 1946 to early 1947. Kilburn attended these lectures [10]. It is clear at least that Newman, Turing and Good were, contrary to the impression which may have been left by Lavington, active in disseminating ideas on computer design and in educating engineers to the point where they could engage constructively with the problem of building a computer. The characterization of the mathematicians as passive enthusiasts runs counter to the available evidence.

Of course, it is one thing to note that mathematicians within Manchester and more widely were active in stimulating an interest in computing per se but quite another to show that they had a tangible effect on the actual design of the Baby. However, there is no reason why anyone should have ever expected mathematicians to be making that kind of contribution. It is one of the shortcomings of Lavington's account that it suggests active involvement by mathematicians in the detailed circuit design of the Manchester Baby is a pre-requisite for their being full partners in the project. This is, of course, very much to see the world from the perspective of the drawing board or the soldering iron and it is important to recognize that what counts as activity is critically dependent on one's point of view. At Bletchley Park, Newman, along with others, took great care to explain to Flowers and the Post Office engineers precisely what was required of the machine which needed to be produced. This would have involved giving an explanation of enough by way of general principle as would be needed to enable the engineers make progress. In the case of Colossus, it would also have been necessary to provide detailed explanation of the precise statistical techniques which the machine was to employ and an explanation of the sort of changes in German encryption techniques to which the machine might need to respond during its lifetime. Clearly, it was only when the engineers understood exactly what was required of them that they would have been in a position to exercise their particular professional skills. In spite of the general direction in which the flow of information moved in this preconstruction phase, it would be a mistake to characterize the early development of the Colossus as involving active mathematicians and passive engineers. It would be more accurate to think in terms of a joint endeavour involving active dialogue. The fact that the skills of a number of different professions had to be brought to bear on the problem is no demonstration that there were two different computing projects at Bletchley Park only one of which, the Flowers project, led to the successful development of a machine. Nor, it must be said, has anyone ever suggested otherwise. Similarly, even were it the case that the Manchester mathematicians had absolutely no involvement in the detailed circuit design of the Baby we would require more evidence to show that there were two different computer building projects in existence only one of which was a success. If Williams had been engaged in an independent attempt to build a computer it should be possible to find confirmation in the form of contemporaneous documents. The civilian effort to develop a computer (or computers) at Manchester was not subject to official secrecy restrictions and there is no reason to suppose that evidence would be hard to find. There are plenty of documents confirming Newman's activity. We can trace his initial interest in developing a computer to his experience of the Colossus. Newman's intention to build a computer at Manchester is confirmed in numerous documents. We can follow in reasonable detail In Williams' case, the documentary evidence of his interest in developing a storage device for use in a computing machine is incontrovertible but there is nothing, that I am aware of which suggests he had any personal interest in developing a computer at Manchester or elsewhere. He is not called on by the University authorities to report on progress independently of Newman's Royal Society funded project. The evidence for an independent project led by Williams is scant and almost entirely non-documentary. Construing Williams to have been engaged on a rival project requires a great deal of contrary evidence to be set to one side and demands a very partial interpretation to be applied to what remains.

Another ground on which Lavington and others have argued that Williams was engaged on an independent computer project revolves around finances. The two project myth treats as very significant a lack of spending on the development of the Baby out of Royal Society funds. The claim that the Baby enjoyed from the outset significant financial support from TRE in contrast to a complete or almost complete lack of financial assistance from Newman and the Royal Society until after the Baby was complete may be termed the financial myth. Lavington sets out the position in a fairly neutral way:

"The Royal Society grant of £35,000 remained substantially intact for several years, eventually providing for the construction of a building to house the University's Ferranti Mark 1 computer in 1951." [11] *Brian Napper, although somewhat strident in tone, does make very clear the important role of the financial myth in supporting the two project myth:*

"There is no question that the "Baby" was Williams' project not Newman's (and effectively funded by the TRE). The confusion is caused because Newman got a grant of £20000 capital + 5 * £3000 per year for wages to build a computer from the Royal Society in 1946. Also the room the Baby was built in was called the Royal Society Computing Machine Laboratory. I won't go into the full debate, but in my mind the empirical proof is in the University records, which show that "Royal Society" was stripped from the name after a year or two, and all the capital and the remaining half of the wages in Newman's grant was spent in 1950 on a new building to house the Ferranti Mark I -the 3rd generation of Williams' (and Kilburn's) computers !!" [12]

There are two substantial defects which run through these accounts. First, they fail correctly to represent how and where the Royal Society Grant was actually spent. Secondly, they take no explicit notice of the details of Newman's bid to the Royal Society or his estimate of expected costs. Thus, the financial myth ignores the financial context. By January 1946, it had been made clear to Newman that while the university approved of his plan to build a computer they were unwilling to provide financial support. Writing to the Royal Society in support of his application on 28th January 1946, Newman gave approximate financial projections, which included provision for £800 for the salary of an Engineer and £500 for two "half-time" mathematicians. No detailed breakdown of costs is given for the construction of the machine beyond noting that the total cost would be £10,000 over the first three years and £20,000 over the first five years. There is no indication that the project was intended to come to an end after the fifth year and in the light of other comments he makes it seems reasonable to suppose Newman saw Manchester as staying at the forefront of computing over the long term. Newman made no allowance for his own salary presumably because felt he could direct the project in the time he had available after his departmental duties were complete. His managerial style, which as we have seen, involved picking very capable lieutenants and giving them freedom to do their jobs without unnecessary interference, would have made this a realistic expectation. Whereas Newman had expected to spend £1300 during the first year of the project to cover salaries he ended up needing only £500. In addition, approximately £300 was spent on sending Rees to the Moore School lectures. Allowing another £50 for miscellaneous spending, this would have brought the first year's actual expenditure up to c.£850. TRE's contribution to the project represented a saving against estimates worth some £500 for Kilburn's salary and a further £100 (approximately) in donated components. It is difficult to place an firm cash value on the hardware TRE supplied as it was mostly surplus stock which if it had appeared on the open market would have had a serious effect on prices Consequently, it would almost certainly have been destroyed if it had not been given to Manchester14. The department of Electro-Technics contributed some test equipment which they built in-house and a quantity of small consumables from their own stores. Additionally they provided infrastructure e.g., floor space, drawing office facilities, workshop facilities. In total we might nominally value this contribution as being worth £150.15

During the first year of the project's life Newman spent around 63% of his estimated budget. The difference between anticipated and actual spending was entirely the result of Kilburn's salary being covered by TRE whose additional generosity in donating components had no further beneficial effect on Newman's first year projections since he hadn't allowed for them in any case. There is no reason to believe that Newman would not have covered Kilburn's wages had the TRE proved reluctant, since he had allowed for such expenditure and had been granted the necessary funds. It is extremely unlikely that the Royal Society, or any other grant awarding body, would have looked kindly on any attempt by Newman to pay Kilburn's salary when alternative funding had unexpectedly appeared. In fact, it should be said that the Royal Society showed great latitude in allowing the £20,000 originally intended for capital development to be spent post hoc on a building. It is also worth mentioning that had the Royal Society instead clawed back the capital grant, matching funds would have had to be found from the budget supporting further computer development at Manchester. The

Royal Society can reasonably be thought of as major investors in the development of the machines that followed the SSEM. However, despite the financial scale and political importance of the contribution made by the Royal Society to the SSEM and to its successors, the part it played has received little acknowledgement from historians.

Newman's spending (Mathematics Dept):
First Year:[1947]

Two half-time mathematicians	£ 500
Travel Expenses (Rees to USA)	£ 300
Office expenses	£ 50
	£ 850

Newman also supplied a huge quantity of components from Bletchley Park (several tons)

TRE spending (estimated):
First Year :[1947]

Kilburn's Salary	c£500
Components (nominal value)	c£100
Infrastructural support (nominal value)	c£150
	c£750

Williams' spending (Electro-technics Dept. estimated):

First Year:[1947]

Office expenses and lab space	c£100
	c£100

In view of the actual spending that took place and the financial context within which Newman was operating, Napper's bald statement of the financial myth, made in the course of commenting on the development of the Baby: "That 'Professor Newman had a grant' was true, but in effect it was not used until it was not required." [13] can be seen to fall very wide of the mark.

8   Conclusion

I have covered a substantial amount of ground. Based on a complete re-examination of the available source material, I have provided a new historical interpretation of the development of one of the most iconic machines in the history of British computing. In doing so, I have called into question the dominant historical narrative and set aside an account of the period which has hitherto remained unchallenged. I have identified three persistent myths that have acted to distort the picture of early computing in Britain and argued, I hope persuasively, that they are without foundation.

In showing that the origins of the Manchester Baby lie in the wartime setting of Bletchley Park and that the mathematicians who worked on Colossus were not only engaged in a process of knowledge and technology transfer were acting with the full knowledge and authority of the authorities I have cleared the way for historians of computing to consider for the first time the extent to the British government played an active part in fostering and directing technological innovation and development in the immediate aftermath of the war.

References

1. Hilbert, D., Ackerman, W.: Grundzüge der theoretischen Logik. Springer, Berlin (1928)
2. Davis, M.: Computability and Unsolvability, 3rd edn. Dover, New York (1982)
3. Evans, C.R.: Interview with Maxwell Herman Alexander Newman (Transcript by D.P. Anderson). Science Museum/National Physical Laboratory (1975) (unpublished interview)
4. On computable numbers, with an application to the Entscheidungsproblem. Proceedings of the London Mathematical Society, Ser. 2 42 (1937)
5. Anderson, D.P.: Was the Manchester 'Baby' conceived at Bletchley Park? BCS eWIC (2008), http://www.bcs.org/upload/pdf/ewic_tur04_paper3.pdf
6. Newman, M.H.A.: Letter to Colonel Wallace, D.D.(A), Government Communications Headquarters, Bletchley Park, August 8 (1945) (unpublished letter)
7. Newman, M.H.A.: Letter to John von Neumann. In: Box 6 Folder 2 Item 2 The Newman Digital Archive, the History of Computing Group and St John's College Cambridge, February 8 (1946a) (unpublished letter)
8. Evans, C.R.: Interview with Frederic Calland Williams (Transcript by David P. Anderson). Science Museum/National Physical Laboratory (1976a) (unpublished interview)
9. Evans, C.R.: Interview with Donald Michie (Transcript by David P. Anderson). Science Museum/National Physical Laboratory (1976b) (unpublished interview)
10. Hodges, A.: Alan Turing: The Enigma. Simon & Shuster, New York (1983)
11. Lavington, S.H.: A History of Manchester Computers, 2nd edn. British Computer Society (1998)
12. Napper, R.B.E.: The Moore School Lectures and the British lead in stored program computer development 1946- 1953 (2004),
    http://www.virtualtravelog.net/entries/2003/10/the_moore_sch ool_lectures_and_the_british_lead_in_stored_program_computer _development_1946_1953.html (retrieved December 11, 2005)
13. Napper, R.B.E.: Newman's Contribution to the Mark 1 Machines (1998),
    http://www.computer50.org/mark1/newman.html (retrieved December 6, 2005)

50 Years Ago We Constructed the First Hungarian Tube Computer, the M-3: Short Stories from the History of the First Hungarian Computer (1957-1960)

Győző Kovács

John von Neumann Computer Society, Hungary,
Former Secretary General and Vice-President
kovacs@mail.datanet.hu

Abstact. The M-3 computer was constructed by members of the Cybernetics Research Group of the Hungarian Academy of Sciences (Hung: Magyar Tudományos Akadémia Kibernetikai Kutató Csoportja, abbr. MTA KKCs) from mid 1957 until the beginning 1959. I was a member of MTA KKCs until 1967. The Group was established for the sole purpose of constructing the first Hungarian electronic tube computer, first the B-1, then the M-3, which began the age of computers in Hungary. We received the basic design of the M-3 computer – mid 1957 - from the Soviet Union, but we received the necessary parts (vaves, cuprox diodes, connectors etc.), too, we used only resisitors and capacitors from a Hungarian plant REMIX. The whole mechanical and electronical consrtruction (logical unites, casing, drum etc.) was done by our mechanical engineers and our mechanical and electronical workshop.

Keywords: Hungary, MTA KKCs, tube computer, first computer-program in Hungary, first computer music, drum, first Hungarian computer export.

1 Introduction: The B-1 Computer

The idea of constructing a computer was born in the Central Prison of Hungary. During the very hard Communist era (mainly the late forties and early fifties) the political leader of the state, the Communist party declared the workers and peasants to be the ruling class. Some Hungarian intellectuals, considered as the enemies of the communist political system, were sent to prison by courts of "justice".

Dr. Rezső Tarján was an insurance mathematician, but in the beginning of the fifties, he became the head of the board of directors in the Ministry of the Industry. He had a lot of personal and official connections with other mathematicians mainly in the Western countries, therefore – in 1953 - the political police constructed „a legal procedure" against him and he was imprisoned. The charge brought against him was: espionage.

During his prison time, he was working in a technical development organisation of the prison, KÖMI 401, together with other two intellectuals: **József Hatvany** physician and **Dr. László Edelényi** mechanical engineer.

A. Tatnall (Ed.): HC 2010, IFIP AICT 325, pp. 68–79, 2010.

They knew about the American and the English computers, therefore they decided, they will construct a Hungarian electronic serial computer, similar to EDVAC /EDSAC. He gave a name to the imagined computer: **B**(udapest)**-1**. They prepared a preliminary study about the future computer, and the director of the KÖMI 401 sent it to the Mathematical Department of the Hungarian Academy of Sciences (HAS). They – naturally - refused it.

The prisoners became free in 1955, the HAS permitted Dr Tarján to follow his research activity of B-1, which was started in the prison. Tarjan could first establish a Computer Department in the Institute for Measuring Technology and Instrument of HAS, a new institution, where several young engieneer and mathematician joined him. Some months later the HAS permitted him to establish an independent institution under the HAS called the Cybernetical Research Group of HAS. (Hung. Abbrev.: MTA KKCs)

Tarjan was not satisfied with the new institution, because a Soviet emigrant electronic engineer, **Sandor Varga** was appointed to be the director. Tarjan got the scientific deputy director position. His colleagues from the Computer department, Hatvany and Dr Edelényi followed him to the MTA KKCs.

Several departments were organised in the frame of MTA KKCs, they started several different developments. The Group did not change their main task to develop the first Hungarian electronic computer, the **B1**. Tarjan employed several new members in the different departments of the MTA KKCs, mathematicians, economists, and engineers. The mathematicians studied the different numerical methods, and the different methods of the programmation. The department of the economists collected a lot of applications and developed programs to solve them etc.

Tarjan employed several new and young engineers. *Fortunatelly, I was one of them.* Our task was to develop basic electronic circuits to the future computer: bistabil and monostabil multivibrators, gates, amplifiers etc. We were very new in the electronic engineer job, (we received our diploma more-less in 1956/57), we learned only the theories and not the practical work, but we started to develop the circuits of the B1.

We had a lot of problems: as I told, we had no practice, neither in the electronic engineering, nor in the computer construction. We hadn't seen any electronic computers earlier, though we could study the first relay computer (MESz-1) constructed by our professor: **Laszlo Kozma** in the Budapest University of Technology. We had none of the necessary knowledge to construct a tube computer, therefore we didn't succeed.

Varga was contrasted with Tarjan, the difference was between them: Varga did not want to develop a computer, his preference was to buy or to construct – as soon as possible – a usable computer. Varga also saw, that we – as young engineers – lacked the practical knowledge to develop and construct a new computer. Varga's first idea was to buy an electronic computer from the Soviet Union, but – in 1956 – the Soviet institutions could not sell computers. One of the first Soviet computer factories in Penza produced the first URAL computers only much later. Therefore this vision was not realistic, yet.

Varga (later Tarjan, too) visited his former Soviet research institution: the Institute of Energetics of Moscow, one of the first institutions constructing the first Soviet "Neumann concept computers". These computers were the Soviet clones of the American IAS computer.

2　The M-3 Computer

During his visit, they developed their first small-medium size computer, the M-3, they started to construct it, too. They offered Varga to give the constructional design of the M-3 computer to MTA KKCs, then we could construct the computer in Budapest.

It is necessary to know that there was an agreement between the Socialist countries, called the "Sofia concept ", because this agreement was born in Sofia. The "Sofia concept" was: the members of the COMECON countries will give – free of charge – their scientific results to the other socialist countries. The M-3 was a scientific result, therefore we received it free!

Varga – and later Tarjan - accepted this opportunity, the documentation of the computer arrived soon – mid 1957 - if I remember well, in two large boxes to Budapest.

We knew that the similar M-3 design was given to the Cybernetical Research Institution of Estonia and the Chinese Academy of Sciences. A little bit later the first M-3 computer was transported from Moscow to Belorussia, Minszk, to the Ordzsonokidze Computer Factory, where they manufactured it in a serial production. In China, they also constructed several M-3 computers, it was the basis of the first Chinese computer production. The design of the M-3 computer was given to Soviet Armenia, too, using this support they constructed their first Armenian computer, the RAZDAN.

The first M-3 computers (in Estonia, China, Hungary and in the Soviet Union) were constructed from the same source, but they were different, because we – in four countries – changed a "little bit" the original designs, and – additionally - we had no connections between the other "M-3" countries. The result was: the M-3 computers – the Soviet, the Estonian, the Chinese and the Hungarian – were not compatible with each other, because we did not harmonise our developments. We could not exchange any software between us, but – during this time – we believed, it was not necessary for anybody else. Everybody wrote their own programs and – generally – did not use the programs of his colleagues, absolutely the institutions used their programs in their countries and not outside of the countries. We did not recognise the importance of the compatibility and the exchange of the software.

Varga reorganised the whole technical department, he replaced Tarján, as the head of our scientific research and appointed **Balint Dömölki**, as the head of the computer development department. I became his deputy, as the responsible head of the technical (electronic) development.

During the construction of the computer my colleagues suggested (me, too) a lot of new solutions, such as: we changed some circuits in the arithmetic unit, we installed some new instructions of the instruction set, we developed a new magnetic drum controller for four drums, we replaced the old input/output devices (Siemens 100 teletype) to fast tape-reader and tape punch equipments etc.

I constructed an amplifier with a loud-speaker, connected to a monostabil multivibrator in the instruction control unit, which – during the computer operation – was oscillated in the voice dominion: between 50 Hz-10kHz. From the beginning of 1959, our computer became an „electronic music instrument", too. A colleague of mine wrote a program, "Beethoven: Für Elise" playing by the M-3 computer.

In the first version of M-3 we used Russian commercial tubes and cuprox diodes, later we decided to replace the Soviet tubes with new long life tubes produced in Hungary. I constructed a totally new control unit for four magnetic drums, I used these new Hungarian Tungsram long-life tubes. I decided to replace the kuprox diodes with Tungsram produced Ge diodes, unfortunately this development was not successful.

We received from the Soviet Union the necessary logical, electronical etc. plans, but we did not get any working documentation of the computer. Balint Dömölki started to study the logical and electronical technical documentation on his own and he understood – step by step - the working method of the computer. He elaborated a new "graphical concept documentation", it was together: a logical and a graphical description, including the circuits of the computer.

At the end we could construct the computer, but we understood the functions of the different parts of the computer from the documentation prepared by Dömölki, not from the original one.

The moment, when the computer – constructed by us – started to work, were very remarkable for us. We all were in the institute – night a day - more-less a week long, we did not go home, we were feeling, (we were sure) the start was very close. Our mathematicians had prepared several programs, they were waiting, too, for the first program to run. Then it happened, but we did not want to believe: our computer was working.

The speed of the first version of our computer was about 30-50 operation per second, our M-3 was in this time the fastest equipment for computing in our country, but it was the largest electronic machine, too. We did not believe that our first program ran successfully, because it was "unbelievably fast"!

Seeing our result, Mr Varga decided to hand the computer officially over to the representatives of the Hungarian Academy of Sciences, but several mathematician members of the Academy did not knew what a computer was.

Varga had an excellent idea: to hand over the computer to one of the best Soviet computer engineers, who designed the M-3 computer. He was **Mr. G. P. Lopato**, chief constructor of the Soviet M-3.

The acceptance test of M-3 was conducted and successfully concluded on January 21, 1959, which was enough to the Presidium of the Hungarian Academy of Sciences. The committee - heading by Mr. Lopato - **declared the M-3 computer ready** and not only the institutes of the Academy but universities, different developers, factories etc. could use it.

After the successful acceptance test, our mathematicians and economists solved several – previously prepared - problems on the new computer not only from scientific domains, but also on the technical and economic life. A number of experts: engineers, economists, mathematicians, linguists and many others started to study computer programming using the machine to solve their own problems.

As soon as in the first few months of the operation of M-3 the KKCs forwarded calculations to the Planning Office solving matrix of the important 5 years economical plans of the socialist planned economy. We had a specialist studying the operation research tasks we also helped with lingual statistics analyses, static calculations for a number of large building constructions. The final control of the statical calculation of the longest bridge - the new Elisabeth Bridge - over the Danube, and many other tasks had been carried out by this machine.

3 The First Education Programs

We organised the first programming courses in Hungary, the participants were mathematicians, engineers, economists and other researchers. The MTA KKCs published the first computer periodical, its title was: "Tájékoztató" (Informatory). Our mathematicians delivered lectures on the new computer programming faculty on the University of Sciences.

In 1960, our colleague **Dr Béla Kreko** suggested and started a new faculty in the University of Economics called "Planing Mathematics". Béla Kreko wanted the University to train very well educated economists, knowledgeable in mathemetics and computer science. Such faculty – I think – was one of the first not only in Hungary and the neighbouring – i.e. socialist – countries, but in all Europe, too. I was invited to organise and teach the computer science in this faculty. The students could study the M-3 computer with some of the ways the programs were running on the computer. I wrote the first university text book on computers. We - Dr Kreko and me - organised the first university computer centre – using an URAL 2 computer - on this university, too (1965).

When the M-3 was successfully tested and accepted by the Hungarian Academy of Sciences (1959), Mr Varga decided – naturally, we supported him – that we would design and construct a new, modern M-3 computer, but he did not ask for a permission from the Presidium of the HAS. We thought, he could not ever get a permission from HAS.

4 The First Computer Centre in Hungary

Mr. Varga reorganised the MTA KKCs, too, he changed the function and the name of the institution, he organised from a research institution the first computer centre in Hungary. This Computer Centre of HAS had several departments, I became the head of the Computer Operations Department. We were working very soon in three shifts, we stopped only in Christmas time. We interrupted the running programs in every 8 hours, 7 hours work and 1 hour maintenance, because the tubes running relatively short time, we had to change several tubes in each shift.

The secret, that we constructed a new, modern M-3 computer, without the permission of the Presidium HAS, became public very soon. The new computer was about 50 % ready. The Academy instructed us to stop our „illegal" work and disassemble the half ready machine. They declared, *the present M-3 is enough for five years long to the institutions of the Academy.*

Varga's penality was, he was kicked out from the computer centre of HAS. The head of the economist's department was appointed as the new director: **Dr Istvan Aczél**.

5 The First Hardware Export from Hungary

Dr Aczél's – as the director – first official trip was to Romania, in 1960, to a computer application conference. He met there two young scientists, **Dr Josef Kaufmann** mathematician and **Viliam Lőwenfeld** electronic engineer, from the University of Timisoara. They informed him, they constructed a computer, its name is **MECIPT-1**, they tried to buy a memory in the Soviet Union, but it was impossible. They asked dr Aczél, to help. Aczél asked me, *„whether we could give a drum to Timisoara?"* I answered: *„Yes, we can, because I prepared in our Computer Centre several reserved drums, if one became defective, we could change it very soon"*. Additionally I gave the drawing of the control unit to them, what I constructed earlier, then we delivered the drum, connected to the computer, which was running without any problems. Naturally: free of charge.

It was the first Hungarian export of computer hardware to abroad. The MECIPT-1 was working till 1968, then they hand over to the Museum Banat in Timisoara.

Unfortunately the MECIPT-1 was not very well accepted by the Ceaucescu political regime. The two designers were Roumanian citizens, but Kaufmann was of Hungarian origin, Lőwenfeld was of German origin, additionally Jewish persons. I was – an original Hungarian – as the third designer (drum). A little bit later, the Museum received a political instruction, they have the MECIPT 1 to kick out from the Museum. Then nobody knows, where was the computer. I was several times in Timisoara, I tried to find the MECIPT 1, but I was not successful.

In 2002, I delivered a lecture in Timisoara then I met a young journalist, **Zoltan Pataki**. I asked him about the MECIPT-1. He asked several other journalists, and we were very fortunate, because we found it in a cellar of the Timisoara Fortress. Then the Alcatel telecommunication company exhibited it, in the same place, in the reconstructed cellar. Whenever I was in Timisoara, I usually visited the MECIPT-1, during at the end of the last year, too. This time I was surprised, because the MECIPT 1 museum was reconstructed, the computer was disassembled and stored in a dirty and unprotected room.

Ceaucescu's Spirit Is Alive

6 The M-3 Was Transported to Szeged, and the End of the M-3 Computer

Returning to the M-3 story, our computer was running 24 hours daily. A lot of users were coming from the different research institutions, universities, but from different workshops, too. They solved a lot of mathematical, economical and technical problems it was a great occasion for the scientific and practical researchers performing their calculations with an electronic computer.

The M-3 computer operated at the Hungarian Academy of Sciences' Computer Centre till 1965, when HAS bought a new URAL 2 (also tube) computer from the Soviet Union. The M-3 computer was transferred to the Cybernetics Laboratory of the József Attila University of Sciences, Szeged, which was headed by Academician **László Kalmár**, Professor of Mathematics and Logic in the University. We had the opportunity to establish the first Computer Centre in the country-side. The head of the University's Computer Centre in Szeged was **Dr Dániel Muszka.**

In 1968 the M-3 became outdated again, the computer was disassembled and the parts of the M-3 were then distributed among the various departments of the university.

The greatest achievement of the development of M-3 was the very early introduction of computer culture to the Hungarian scientific and research community. The M-3 was and still is the symbol for the beginning of the age of computers in Hungary.

7   Technical Characteristics of M-3

7.1   Arithmetic Unit

31 bits/word, parallel computing, four registers, operational speed: addition: 60 microsec, substraction: 70-120 microsec, multiplication: 1.9 millisec, division: 2,0 millisec.

7.2   Input/Output Device

First: Siemens T-100 teletype, tape reader and puncher, 5 position telex code, input/output speed: 7 chrs/sec.

Later the input device was a Ferranti photoelectric tape-reader, 8 position code, speed: 300 chrs/sec, the output device was a Creed puncher, 8 position code, speed: 100 chrs/sec.

7.3   Memory

First a magnetic drum, 1 kWord (later: 1,6 kWord) capacity,

Later as back ground memory... two - simultaneously - running drums were operating together (2x1600 Words), then, as operating memory: a ferrit core memory, its capacity was: 1 kWord.

7.4   Control Unit

Two address code, 31 bits per instructions, 1 sign bit, 6 bits for operational code, 12 bits first address, 12 bits second address.

7.5   Power Supply

Total power dissipation: about 10-15 kW.

7.6   Parts Used (Approxomate Numbers)

About 500 logic units, about 1000 vacuum tubes, about 5000 cuprox diodes, about 4000 resistors, about 3000 capacitors.

Fig. 1. Original photos of the M-3 computer. The only documentation of the successful acceptance test in the daily newspaper – 21. January. 1959. Wednesday - Esti Hírlap: the first Hungarian electronic computer - the M-3 - is ready. The engineering Group: from left to right: S. Pohradszky – later: A. Röhrich, – I. Ábrahám, I. Molnár, L. Szanyi, Gy. Kovács, Zs. Várkonyi, B. Dömölki (in the shadow: K.Kardos).

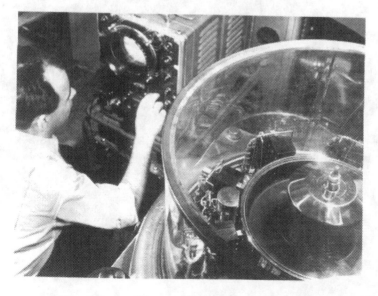

Fig. 2. Gy. Kovacs and the drum memory of the M-3

Fig. 3. The M-3 computer is ready

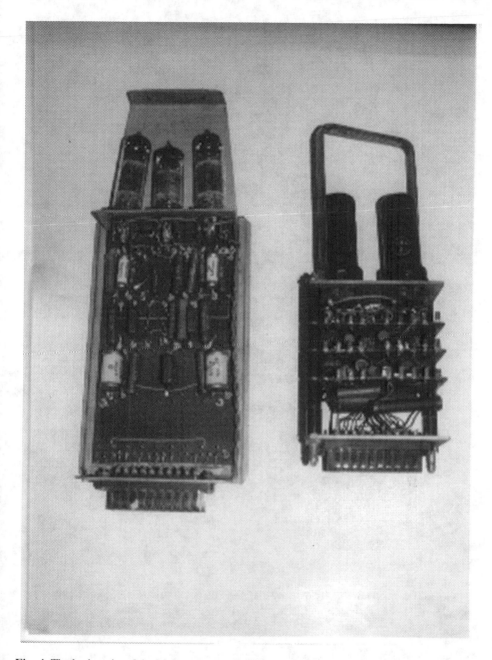

Fig. 4. The logic units of the M-3 computer, (left) the new developed drum controller-unit, with Hungarian produced long-life tubes, (right) the old unit, original Soviet design with Soviet elements

Fig. 5. I found the MECIPT-1 with our drum in a cellar of the Timisoara Fortress (2002)

Acknowledgements

Thanks to **Noemi Adorjan (Canada),** that she corrected my linguistic mistakes in my paper, and **Róbert Horvath**, preparing the camera ready pdf version.

References

1. 40 éves a magyar számítástechnika és a Volán Elektronika 1999. (40 Years Old the Hungarian Computer Science and the Volan Electronics) (1999) (Hung)
2. Számítógéptechnika '68. Az Automatizálási, Információfeldolgozási és Operációkutatási Tanács (AIOT) esztergomi konferenciájának a kiadványa. 1968. (Computer-technology '68. The Proceedigs of the AIOT Conference, Esztergom) (1968) (Hung)

3. Kovács, G.: A számítógépek technikája. (The Technology of the Computers). Tankönyvkiadó (1974) (Hung)
4. Tarján, R.: Gondolkodó gépek (Thinking Machines). Bibliotheca kiadó (1958) (Hung)
5. Dömölki, B., és Drasny, J.: M-3 elektronikus számítógép ellenőrző programjai (The Control Programs of the M-3 Computer) (1965) (Hung)
6. Kovács, G.: Válogatott kalandozásaim Informatikában (My Selected Adventures in Informatics). MASSI és GÁMA-GEO Kiadó (2002) (Hung)
7. Muszka, D.: Szemelvények a számítástechnika szegedi történetéből (Selected Passages from the History of Computer Science in Szeged), Kézirat (1995) (Hung)
8. Sándor, F., Dömölki, B., Révész, E., Szelezsán, J., Veidinger, L.: Az M-3 elektronikus számológép programozása (The Programation of the M-3 Computer), kézirat (1958) (Hung)
9. Szentiványi, T.: A számítástechnika kezdetei Magyarországon (The Beginning of the Computer Science in Hungary). Természet Világa, vol. 125 (1994) (Hung)
10. Kovács, G.: The History of M-3, the first Hungarian Electronic Digital Tube Computer. In: Proceedings of Workshop of MEDICHI 2007 (Methodic and Didactic Challenges of the History of Informatics) (2007)

Anatoly Kitov - Pioneer of Russian Informatics

Vladimir A. Kitov[1] and Valery V. Shilov[2]

[1] Fujitsu Technology Solutions, Zemlyanoi val 9,
105064 Moscow, Russia
{vladimir.kitov,olga.kitova}@mail.ru
[2] "MATI" – Russian State Technological University, Orshanskaja 3,
121522 Moscow, Russia
shilov@mati.ru

Abstract. Anatoly Ivanovich Kitov (1920-2005) is one of the most outstanding representatives of the first generation of scientists who had created Russian cybernetics, computer engineering and informatics. Due to political reasons his many-sided scientific and organizational activities were hushed up. Only recently publications adequately estimating his role appeared. The report represents biography and summary of main scientific achievements of A.I. Kitov as well as short description of his scientific school and his collaboration with IFIP and other international organizations.

Keywords: Anatoly Kitov, cybernetics, informatics, programming languages, management information systems, Computer Centre № 1, M-100 computer, Red Book project, computer-aided control systems, TC-4 IFIP, MEDINFO, IMIA.

1 Introduction

This report is about Anatoly Ivanovich Kitov - pioneer of cybernetics, informatics and management information systems (MIS) in the USSR and Russia.

Anatoly Kitov was brightest outstanding personality who devoted his life to development of cybernetics, computer and software engineering, programming languages, information retrieval systems and management information systems as well as to their practical implementation in various fields of human activity: national economy, military defense, health care and medicine.

A. Kitov was the real pioneer, words "the first" and "for the first time" can be applied to all stages of his scientific career. In the USSR A. Kitov is the author of the first positive article about cybernetics, the first thesis on programming, the first Soviet book about computers and programming, the first research and articles on non-arithmetic usage of computers, the first project of all-national computer network, the first national textbook on computer science, the first scientific report on management information systems (IMS), etc. He created the most powerful in the USSR computer at his time, the Computer Centre № 1, associative programming theory, standard industrial management information system (MIS) (in the Ministry for Radio Engineering Industry), the first Soviet management information systems for health care, two new high-level programming languages (ALGEM - for solving economic tasks

A. Tatnall (Ed.): HC 2010, IFIP AICT 325, pp. 80–88, 2010.

and NORMIN - for working with text information), etc. The total amount and innovative quality of his scientific works are really impressive.

The Chairman of IFIP Congress 1974 and IEEE Computer Society Computer Pioneer academician Victor Glushkov stressed the outstanding role of A. Kitov: "The real implementation of cybernetical ideas in the USSR takes start from the Kitov's first positive article "Main features of cybernetics". Kitov's book "Digital Computing Machines" was the first Soviet book on this topic".

Alexey Lyapunov, also the IEEE Computer Society Computer Pioneer, called A. Kitov "the first knight of Soviet cybernetics".

2   The Beginning: World War II and Learning at the University

Anatoly Ivanovich Kitov was born on the 9th of August 1920, in town of Samara in Volga region. During the Civil War (1918-1920) his father Ivan Stepanovich Kitov served in White army as a junior officer what could have negative consequences those times. So in 1921 Kitov the elder, with wife and son, moved to Tashkent (Uzbekistan). Anatoly Kitov graduated from secondary school in 1939. For excellent results in study and brilliant abilities he was awarded a gold medal. In the same year he entered the Tashkent University but after three months was called up for military service. Since he again demonstrated outstanding talents the army commander in chief field marshal Kliment Voroshilov personally ordered to enlist Kitov into High Artillery School in Leningrad. With outbreak of the war with Germany, in July 1941, Anatoly was dispatched to the South front as an artillery platoon commander. In the battle for Stalingrad Kitov was seriously wounded. He spent all four war years in anti-aircraft artillery, where he himself steadily continued study of mathematics and physics using every free minute.

With the end of war, in 1945, A. Kitov entered the Artillery Engineering Academy in Moscow. Formidable knowledge obtained during his excellent studies enabled Kitov's beginning of own scientific research already before graduation. Being still a student he received a patent (the USSR "Author's Certificate") for devised by him "jet cannon" and the Ministry of Defense reported about it directly to Joseph Stalin. A. Kitov graduated from the Artillery Engineering Academy in 1950 with the Gold medal.

In 1952 Kitov successfully submitted dissertation thesis "Programming for ballistic problems of the long-range rockets" and received scientific degree "Candidate of technical sciences". It was the first Soviet thesis on programming. In 1952-54 he created and headed the first computer department at the Artillery Engineering academy.

The task of the national importance was carried out by A. Kitov in 1954 when he created and became a chief of the Computer Centre № 1 of the USSR Ministry of Defense. Kitov managed to form the team of scientists and engineers, to teach his colleagues, he founded the main directions of scientific research of this pioneer computer center. In two years the Computer Centre № 1 became the most important scientific research and engineering center in the USSR. In the middle 1950-s 160 programmers, 85 information analysts, 40 mathematicians (specialists in mathematical modeling), several hundreds specialists in computer engineering worked on specialized and universal computer projects.

Under the Kitov's leadership in Computer Centre № 1 such important tasks were solved as: calculation of orbits of all Soviet space stations and artificial satellites of the Earth, including space flights of Yu. Gagarin and other first Soviet astronauts; creating specialized computers for military use, tasks of the Soviet Army including special tasks of Main Artillery Department, Main Intelligence Department, Strategic Missile Department, modeling different military combat situations with the use of tanks, airplanes and artillery, working with large amounts of information, and many others. In the Computer Centre № 1 Kitov created departments which worked out central processors, computer memory and storage for new computers, department of mathematical linguistics, etc. The most important research included theoretical programming issues and working out complex computer systems with unique software. In the Computer Centre № 1 Kitov regularly held scientific conferences and seminars, edited the scientific journal. The well-known Soviet scientist colonel Vladimir Isaev remembers: "If we make parallels with, very popular then, space researches, Kitov was in position some half-way in-between academician Mstislav Keldysh (leading theorist) and academician Sergey Korolev (main designer of Soviet spaceships). That's my personal opinion. At the Computer Centre № 1 he was the most initiative one, always generated ideas and formulated problems. So, generally speaking, he was the brain of the projects".

In the 1950-s the role of this top secret Computer Centre for the development of Soviet computer science was difficult to overestimate. Taking into account the variety and complexity of scientific research, number of qualified specialists it was the largest and the most powerful scientific computer center in the USSR and one of the best in the world.

Fig. 1. Colonel Anatoly Kitov, scientific head of Computer Centre № 1 of the USSR Ministry of Defense (1959)

3 Scientific Results and Achievements

The most important role in Kitov's life played the book by Norbert Wiener "Cybernetics or Control and Communication in Animal and Machine" which Anatoly Ivanovich read in 1952 in the secret Soviet library. Communist authorities of the USSR under the guidance of Joseph Stalin at that time named cybernetics "pseudo-science and servant of imperialism". Anatoly Kitov needed great courage not only to read the Wiener's book but to write article "Main Features of Cybernetics" - the first positive article on cybernetics in the USSR. In the Stalin times he risked not only by his freedom but also by his life. During two years, since 1953 till 1955 A. Kitov delivered a lot of lectures about cybernetics, but this article was published only in 1955 after long negotiations with Soviet ideologists after Stalin death. A. Kitov asked mathematicians Sergey Sobolev and Alexey Lyapunov to be co-authors of the article. This article became the starting point of Soviet cybernetics. Academician Guriy Marchuk, former president of the Academy of Sciences of the USSR, wrote: "the article made decisive impact on the academic public, which obtained a firm ground of new science, stimulating the people to change and update their mentality. In the article the authors considered general scientific meaning of the cybernetics as emerging theory of information science, as well as the theory of electronic computing machines and systems of automatic control. That was really the scientific event of great value".

Since 1953 till 1960 A. Kitov published several fundamental scientific works on informatics and cybernetics. In January 1956 Kitov published his book "Electronic Digital machines" - the first Soviet book about computers. The last part of it was about using computers in economics, production processes automation, artificial intelligence, etc. He understood that computers were able to transform all spheres of human life. American scientist John W. Carr who analyzed more than 150 main publications on the topic wrote about Kitov's book in his book "Lectures of Programming" (1958): "Probably the most comprehensive description of the computer programming problems, illustrated with detailed analysed examples of both manual and automatic programming, is available at present in the book by A. Kitov. Some parts of it have been translated into English so they are available at the American Computer Association".

So, in the 1950s, the Kitov's book was really one of the best books dedicated to computers and programming.

"Electronic Digital machines" by Kitov was published in the USA, China, Czechoslovakia, Poland, German Democratic Republic and other countries. Guriy Marchuk wrote: " Electronic Digital machines" by A. Kitov, published in 1956, was the first systematized course for a broad circle of students and specialists who began mastering electronic computers, computing and their practical applications. The book actually caused an overturn in attitudes of many researchers since it had been written simply and easily, clearly presenting comprehensive amount of well systematized material".

In 1956 one more book "Elements of Programming" by A. Kitov and his two co-authors was published.

In 1956-1957 A. Kitov started in the Ministry of Defense and in the USSR the new scientific research area "Development and Usage of Information Retrieval Systems".

In 1958 A. Kitov published brochure "Digital computing machines" in 42,000 copies in which the perspective of comprehensive automation of administrative work and management in the country was described for the first time. He suggested to connect all computing centers into one national united network.

In 1958 Kitov, in collaboration with colleagues received patent for new operation principle of computer central arithmetic device, so-called "Method of computer command rate quadruple combination". That method was implemented in computer M-100, the most powerful Soviet computer at that period of time. It was used in solving problems of antiaircraft missile aiming at flying target and in other military tasks. A. Kitov was the chief designer of M-100. Its architecture was described in his dissertation thesis "Computer applications for problems of antiaircraft defense", which was successfully submitted in 1963 and was awarded "Doctor of technical sciences" degree. He analyzed basic principles of special computers design and structure, their programming languages, as well as mathematical computer modeling of dynamic systems for antiaircraft defense problems such as ballistics, flying target tracking, etc.

In 1958 A. Kitov and N. Krinitskiy published the book "Electronic Computers". In 1962 Pergamon Press published English translation of this book; in the preface professor A.D. Booth wrote: "This book gives, for the first time in the West, the Russian approach to an elementary description of the principles, construction and programming of automatic digital computers. It will be noted that special reference is made to the solution of mathematical and logical problems and to the automatic control of processes. It is particularly interesting for the worker in this field to see how closely this treatment follows that which has developed in the West, and to observe the modifications which have resulted from the computing machines which are available in Russia. Electrical and Electronic engineers, mathematicians, physicians and all concerned with the design and use of computers will welcome this Russian work".

Fig. 2. The book "Electronic Computers" by A. Kitov and N. Krinitskiy in Russian (1958) and English (1962)

In 1959 A. Kitov and N. Krinitskiy published large textbook "Electronic Digital Computers and Programming" which became the most popular one among engineering universities. Several generations of students in the half of the world (in the USSR, Eastern Europe and China) studied with that book which was one of the best in the field of programming and computers at that time.

It is necessary to mark out two pioneer initiatives of A.Kitov which had national importance. In January 1959 A. Kitov sent a letter to the USSR leader Nikita Khrushchev "About creating automated management system for national economy" where he proposed to create national computer network to be used for multiple purposes, first of all to manage Soviet economy. Anatoly Ivanovich added his brochure "Digital computing machines" (1958) to this letter.

The top leaders of the country partly supported Kitov's initiatives, and special government resolution was approved with the decision to produce modern computers and use them in production automation. But main ideas of Kitov's letter regarding managing economy with the help of national computer network were not approved.

In November 1959 A. Kitov made the first in the USSR report on management information systems (MIS) for enterprises and industries. In autumn 1959 A. Kitov sent the second letter to Nikita Khrushchev in which he suggested the way to minimize expenses of creating national computer network. He added to this letter project in 200 pages (called Red Book) of creation unified automated administrative control system to be used simultaneously for army and civil economy. It should be based on common network of computer centers established and maintained by the Ministry of Defense. Concentration of computers in powerful centers with reliable maintenance, run by military personnel, would sufficiently raise quality of their usage.

The second Kitov's initiative had worse consequences than his previous letter. He criticized the situation with using computers in the USSR, especially in the Soviet Army, and it caused discontent and rage of high authorities. The project was rejected, and A. Kitov was expelled from the Communist Party of the USSR, moved off his administrative position and left the Soviet Army.

It was a severe blow. That could be certain end of any political and scientific career for most of those who would, by a misplay of luck, find himself in similar situation but not for A. Kitov. He was a man of ideas and real patriot of his country. Although being forced to start again with another work, in the beginning of the 1960-s, he did not quit the idea of global automatic control system. In 1961 he published one of his main works in the field of management information systems - "Cybernetics and management of national economy". Kitov considered Soviet economy as a complex cybernetical system which had to be optimized. To manage it efficiently it was necessary to build distributed national network of computing centers which would work with economic information. He suggested to join main computers into United centralized management information system for national economy. This article by was highly appreciated by Soviet and foreign specialists, including specialists from the USA. The big positive review on this article appeared in 1963 in the *Operations Research* journal (Vol. 11, № 6, November-December).

In 1962 Kitov made a report called "Associative programming" in which stated the main ideas of his theory of working with large information units. He authored and introduced notion "associative programming", defining it as following. "Associative programming is a totality of solution methods, intended for information logic problems, which are based on programming of associative relations between the data stored in computer memory … In other countries this range of problems is also called: list-processing, node-processing, chain-addressing, control words method, etc."

In the middle of 1960th A. Kitov was appointed the Chief designer of the Industrial management information system (MIS) of the Ministry for Radio Engineering

Industry. His team worked out main algorithms for MIS, created modeling methods and produced sufficient amount of software. That system was highly evaluated by the governmental experts and recommended as standard for other ministries connected with defense production. Academician Viktor Glushkov, director of Institute for Cybernetics and vice-president of the Academy of Sciences of Ukraine, was scientific leader of the MIS project in the USSR. Kitov kept close partnership with him.

In 1967 A. Kitov published his next fundamental book "Programming for information-logical problems" about information retrieval and management information systems. This Kitov's book was translated into German. Serious Kitov's achievement was creating a new programming language ALGEM in the middle of 1960-s to program economic problems. ALGEM was implemented and used in hundreds of enterprises in the USSR and East European countries. In 1971 Kitov published fundamental book "Programming for Economic and Management Problems" (400 pages).

In the 1970-s Kitov turned to implementation of information systems and computer engineering in medicine and health care. That was the period when automatic systems for control and management became very popular. Those years Kitov performed design of management information system "Health care". He formed information model of medical branch, developed standard structure of the system, software packages for information arrays control, developed logic, structure and functional algorithms for information retrieval systems, etc. His principal monograph (1976) written about that project was named "Automation of Information Processing and Control in Health Care". In 1977 publishing house "Medicine" produced his new book, "Introduction to Medical Cybernetics", and in 1983 one more on the subject, called "Medical Cybernetics".

Generally speaking his activity with medical information systems was much nearer to contemporary issues than one could judge from the titles. For example he managed to install in a Moscow hospital one of the first PDP-11/70 - a highly efficient mini-computer of the middle-1970-s. Its programming system MUMPS - Massachusetts General Hospital Multiprogramming System (later it was standard ISO11756:1991) was popular in the USSR as programming system DIAMS for minicomputers SM-4 (similar to PDP). It was predecessor of modern M-technologies for medical applications, supported by post-relational database control system Cache of Inter-Systems.

Kitov is famous as one of the leading scientists in the field of information retrieval systems (IRS), algorithmic languages and methods of associative programming. Results of his researches were presented in his monographs, "Programming for information-logical problems" (1967) and "Programming for Economic and Management problems" (1971). Kitov leaded the group of specialists who created new general programming language NORMIN to work with normalized text information. It was implemented in many Soviet health care organizations. Kitov used methods of cybernetics, system analysis, etc. to solve medical and health care problems in the situation of risk and lack of information. Three fundamental books, articles and scientific reports describing computer systems implemented in medical and health care organizations created the foundation of medical cybernetics and informatics in the USSR.

Anatoly Kitov made notable contribution as academic teacher. He supervised and consulted Candidate and Doctoral dissertations of more than 40 scientists from Russia, Ukraine, Uzbekistan, Latvia, Moldova, Germany, Hungary, Bulgaria, Poland, China, Viet-Nam and other countries. He also has been a member of the Russian "Programmirovanie" journal editorial board from the very day of its establishing.

Since 1980 till 1997 A. Kitov was professor of Plekhanov Russian Academy of Economics, head of the Computer science chair. He is the author of 12 monographs in computer science which are translated into 9 foreign languages: English, German, Japanese, Chinese, Polish, Hungarian, Romanian, Bulgarian, Czech. His work opened for several generations of specialists the wonderful world of cybernetics and information technologies, founded in the USSR and Russia military, economic and medical informatics.

Fig. 3. Participants of the International conference of Medical Informatics at Japan (1978). A. Kitov is the fifth in the second row.

4   Anatoly Kitov and IFIP

During more than 12 years A. Kitov was involved to the IFIP activity. In MedINFO A. Kitov had status "The national representative from the USSR". He was the member of Technical Committee № 4 (TC-4) IFIP. A. Kitov had regular contacts with chairman of TC-4 IFIP professor Jan Roukens (Netherlands) and vice-chairman professor B. Schneider (Germany). A. Kitov took part in three International MEDINFO Congresses:

- MEDINFO-1974, Stokholm (Sweeden).

- MEDINFO-1977, which was held in August 1977 in Toronto (Canada). On this international forum A. Kitov was the chairman of Session T2 - "Biomedical Research General".

- MEDINFO-1980, which was held in September 1980 in Tokio (Japan). Here A. Kitov was the Member of the Programming Committee. Also A. Kitov was IMEA officer from the USSR.

A. Kitov took active part in the work of international committee TC-4 IFIP and in other events abroad, e.g. TC-4 IFIP session in Dijon (France), TC-4 IFIP session in Florence (Italy), TC-4 IFIP session in Amsterdam (Netherlands), in 1978 he took part in the MEDIS'78 in Japan, which was held in Tokio and Osaka, etc.

Anatoly Ivanovich Kitov died on the 14th of October, 2005 in Moscow.

References

1. Gerovitch, S.: InterNyet: why the Soviet Union did not build a nationwide computer network. History and Technology 24(4), 335–350 (2008)
2. Gerovitch, S.: Mathematical Machines of the Cold War: Soviet Computing, American Cybernetics and Ideological Disputes in the Early 1950s. Social Studies of Science 31, 253–287 (2001)
3. Gerovitch, S.: Russian Scandals: Soviet Readings of American Cybernetics in the Early Years of the Cold War. Russian Review 60, 545–568 (2001)
4. Neskoromny, V.: Chelovek, kotoryi vynes kibernetiku iz sekretnoi biblioteki (Человек, который вынес кибернетику из секретной библиотеки). Komp'yuterra 43, 44–45 (1996)
5. Chernyak, L.G.: Anatoly Ivanovich Kitov – inzhener i myslitel' (Анатолий Иванович Китов - инженер и мыслитель). PC Week/RE 43 (1999)
6. Ruzaikin, G.I.: Pamyati Anatoliya Ivanovicha Kitova (Памяти Анатолия Ивановича Китова). Mir PK 2, 82–83 (2006)
7. Mironov, G.A.: Pervy VC i ego osnovatel' (Первый ВЦ и его основатель). Otkrytye Sistemy 5, 76–79 (2008)
8. Isaev, V.P.: Ot atoma do kosmosa: 50 let ASU (От атома до космоса: 50 лет АСУ). Otkrytye Sistemy 5, 57–59 (2009)
9. Kurbakov, K.I.: A.I. Kitov – odin iz osnovopolozhnikov otechestvennoi kibernetiki (А.И. Китов - один из основоположников отечественной кибернетики). In: Kibernetika - ozhidaniya i rezul'taty, vyp. 2, Znanie, Moscow, pp. 40–44 (2002)
10. Malinovsky, B.N.: Istoriya vychislitel'noi tekhniki v litsakh (История вычислительной техники в лицах). KIT, Kiev (1995)
11. Dolgov, V.A., Shilov, V.V.: Ledokol. Stranitsy biografii Anatoliya Ivanovicha Kitova (Ледокол. Страницы биографии Анатолия Ивановича Китов). Novye technologii, Moscow (2009)
12. Dolgov, V.A.: Kitov Anatoly Ivanovich - pioneer kibernetiki, informatiki I avtomatizirovannykh system upravleniya (Китов Анатолий Иванович - пионер кибернетики, информатики и автоматизированных систем управления). Plekhanov Russian Academy of Economics, Moscow (2009)
13. Gerovitch, S.: From Newspeak to Cyberspeak. In: A History of Soviet Cybernetics. The MIT Press, Cambridge (2002)

Materiel Command and the Materiality of Commands: An Historical Examination of the US Air Force, Control Data Corporation, and the Advanced Logistics System

Jeffrey R. Yost

Charles Babbage Institute, University of Minnesota

Abstract. In the late 1960s the US Air Force Logistics Command (AFLC) engaged in an unparalleled, real-time computer networking project to manage all its logistics (location, inventory, maintenance, and transportation of personnel, aircraft, weapons, components, spare parts, etc.), the Advanced Logistics System (ALS). The $250 million ALS project was substantially larger in size and cost than earlier real-time computer networking projects (including SAGE programming and SABRE), but it has received virtually no attention from historians of computing. Ultimately, the ALS project failed. Drawing from an oral history with lead contractor Control Data's (CDC) longtime ALS project manager, previously unavailable CDC documents, and documentation and an oral history from a leading external Air Force advisor on ALS, it shows how the AFLC pushed too far too fast in seeking to be a first-mover in creating a massive unified database and real-time computer network for highly complex logistics.

Keywords: Air Force Materiel Command (AFMC), Control Data Corporation (CDC), Advanced Logistics System (ALS), supply management, technological failure, real-time computing, and computer networking.

1 Early Air Force Logistics and Information Technology Applications

The importance of Air Force logistics (managing the location, transportation, maintenance, and supply of personnel, aircraft, weapons, components, spare parts, etc.) to military operational effectiveness and cost containment is impossible to overemphasize.[1] The trade journal *Business Machines* reported that the Air Force Logistics Command in 1960 managed more assets than any organization in the world—more than General Motors, United States Steel, Metropolitan Life, and Western Electric combined [1]. Information systems and communication technologies have long been central to aiding Air Force logistics. In 1926 the Air Force Materiel Division

---

[1] Some weapons, or weapons systems, were managed by the Air Force Services Command (AFSC)—see footnote 2.

A. Tatnall (Ed.): HC 2010, IFIP AICT 325, pp. 89–100, 2010.

installed its first punched card tabulator, at McCook Field in Dayton, Ohio [2]. Dayton has continued to be the home of Air Force logistics management for more than 80 years—under the name Air Materiel Command (AMC) up to the early 1960s, Air Force Logistics Command (AFLC) from the early 1960s to the early 1990s, and Air Force Materiel Command (AFMC) from the early 1990s to the present (all at Wright-Patterson Air Force Base from the late 1940s forward) [3].[2] Following this initial IT procurement in the mid-1920s, the AMC installed numerous mechanical and electromechanical tabulators during the inter-war years and throughout World War II. In July 1954 the AMC was an early adopter of digital computers, installing a UNIVAC I, which was soon followed by IBM 650s and 705s. Interestingly, in 1956, it first used an IBM 705 to manage personnel much like other inventory—a practice/plan that continued with ALS [1]. Without much central organization, the other five Air Materiel Areas (AMA)[4][3] also procured digital computers in the 1950s [5]. At this time, the data processing task of the Air Force's logistics headquarters and other AMAs were managed individually, and the data processing technology was very much distinct from the communication technology used to share logistics information.

At the end of the 1950s and the start of the 1960s the Air Materiel/Logistics Command engaged in efforts to achieve centralized authority over logistics. By the mid-1960s this also included efforts and planning to bring together data processing (digital computers) and communications. The task was daunting as the AFLC had 376 individual information systems—tracking/managing procurement, inventory, transportation, and maintenance—at headquarters and the other AMAs [6]. In late 1966 this effort was formalized in early planning for a massive automated system—a centralized database and network of computers—to provide real-time information to authorized personnel at different AMAs and command posts [7]. AFLC managers and data processing personnel, with the aid of consultants from the COMRESS Corporation, completed a "Master Plan" for the Advanced Logistics System (ALS) in March 1968 [7].

2   An Examination of ALS and Its Failure

The following paper is a short history of ALS. Despite ALS being a larger and far more expensive real-time computer networking project than the frequently examined SAGE programming effort or the development of SABRE, it has been completely ignored in the existing computer history literature.[4] A self-published AFLC history of

---

[2] In 1961, with the renaming of AMC as AFLC, the research and development and weapon system acquisition division of AMC was broken out as the Air Force System Command (AFSC). In early 1992, AFLC and AFSC were combined to form the AFMC.

[3] In addition to the headquarters for logistics at Wright-Patterson Air Base, Ohio, the other five AMAs were: Tinker Air Force Base, Oklahoma; Hill Air Force Base, Utah; Kelly Air Force Base, Texas; McClellan Air Force Base, California; and Robins Air Force Base, Georgia.

[4] Many scholars and writers have examined Semi-Automatic Ground Environment (SAGE—a computer and radar air defense system) programming and the development of the Semi-Automatic Business Research Environment (SABRE—a pioneering airline reservations system). Some of the best source publications and scholarly analyses include [8],[9],[10],[11],[12].

ALS [2]⁵ uses Air Force documents almost exclusively, and places much of the blame for the project's shortcomings with lead computer contractor Control Data Corporation (CDC) [2].[6] My paper takes advantage of a host of new resources, and seeks to provide a more nuanced and balanced interpretation as to why this project failed. It also places this important story within the broader framework of the history of computing and history of technology.[7]

Air Force historian William Elliott, author of the AFLC published history of ALS, presents strong approval of the ALS plan by the Air Force, coupled with brief mention of a few prominent skeptics. Throughout, he stresses that the Air Force trusted the experts—this included outside advisors, consultants from the computer services industry (COMRESS and Computer Sciences Corporation), and computer firms looking to bid for the primary ALS contract [2].

Elliott mentions the RAND Corporation's early experience in computer time-sharing and a presentation by RAND to the Air Force on this topic in 1966 [2]. However, he neglects to discuss RAND's role as a longtime top IT advisor to the Air Force. The RAND Corporation, spun off from the Air Force's Project RAND in the early post-World War II era, was a prominent advisor to the Air Force on many scientific, technical, military, and strategic matters from the late 1940s through the 1970s. While the RAND Corporation, during the 1960s, shifted from a near exclusive focus on military research and development to include a broader social and economic research agenda, it continued to be a top advisor to the military, and especially the Air Force, throughout that decade and beyond.[8] By the start of the ALS project, RAND had conducted pioneering research on inventory management for nearly a decade [15]. The Computer Science (CS) Department (and earlier, the Mathematics Department) at RAND advised the Air Force on computing and software systems throughout the 1950s, 1960s, and 1970s.[9] RAND's head of CS, Willis Ware, led a RAND advisory committee on ALS. In reflecting on the advice he and his committee provided to Air Force leaders prior to the project and in its early stages, Ware stated:

---

[5] There are only several publicly available copies of this history worldwide.

[6] Elliott, while acknowledging shortcomings and failed goals with the project (attributing most of these to Control Data Corporation's inexperience and mistakes), does not present ALS overall as a failure. At times, he in fact stresses the benefits of ALS—emphasizing the importance of the Air Force's early commitment to massive computer infrastructure to future logistics efforts. He, however, generally neglects to provide concrete evidence of direct substantive connections. Nor does he explore what alternatives might have looked like to expand IT infrastructure for Air Force logistics outside of the ALS project.

[7] These new resources include an oral history interview I conducted with the ALS project manager for Control Data (Fred Laccabue); CDC documents, including correspondence with government officials; and an oral history I conducted and documents from the leading external Air Force computer advisor of the time, RAND computer scientist Willis Ware.

[8] For the early history of the RAND Corporation, its broadening scope in the mid-1960s, and its continuing relationship with the Air Force see [13]. A more popular history of RAND covers some similar ground [14].

[9] Prior to the 1970s, computing research and expertise resided within the RAND's Mathematics Department.

The Air Force did have a big problem to retrofit the logistics institution with an advanced computer system. Unfortunately the primary message that we delivered to the Air Force at the time was, "whatever you think you're doing, you're doing it poorly and you're not organized to do it well." Which is not a pleasant message to deliver, but the fact of the matter is that Air Force at that juncture...had never faced or managed the conversion of a huge computer system. And this was a huge operation [16].

Ware's correspondence and reports in the RAND Corporate archives fully corroborate these comments. Repeatedly, RAND computer scientists conveyed to the Air Force the great risks of implementing such a massive, real-time, third generation computer networked system—with which the Air Force had no prior experience. As the project moved forward, Ware and other RAND computer scientists provided advice regarding specific areas—such as file conversion, performance analysis and test simulation, and computer security—to improve the likelihood for success with ALS, but did not waiver from their assessment of the major risks involved.[10] While COMRESS and Computer Science Corporation, two computer services firms serving as pre-project consultants, undoubtedly presented less dire assessments,[11] no organization (given JOSS—RAND's pioneering time-sharing system—and RAND's connection with its spin-off, System Development Corporation) was better equipped to advise on a massive, pioneering real-time computer and software project than RAND.[12]

Prior to the launch of the ALS project, the House Appropriations Committee asked the General Accounting Office (GAO) to produce an assessment report on the Air Force's ALS plans. The report cited a number of potential difficulties [4]. Specifically, it reported that "there are strong indications that problems may be encountered in obtaining and implementing computer software."[4] The GAO cited a recent major airline's canceling of a $56 million contract with a computer vendor for a large-scale, multifunction data processing system (less ambitious than ALS)—resulting from recurrent delays and problems with software implementation and data security [4]. The AFLC leaders did take some suggestions, such as implementing a pre-bid benchmark test, but they did not seem to take to heart the direct warnings regarding software challenges [4]. While Elliott cites this report, he only uses it for the Air Force data and projections it presents, or to record early project delays, not to convey the GAO's reservations regarding ALS software [2].

Following a Request for Proposals (RFP) for ALS, meetings in spring and summer 1968 were held at Hanscomb Field (near Bedford, Massachusetts) with firms likely to bid for the project's primary computer contract.[13] Elliott highlights the enthusiastic reception to the RFP by the computer industry, yet one of Control Data Corporation's

---

[10] This is based on examining numerous reports and extensive correspondence files of Willis Ware while conducting research at the RAND Corporate Archives in 2004.

[11] There were current and potential future financial incentives for these software consultancies to favor continued exploration and development of what became ALS.

[12] The RAND Corporation's early pioneering work and considerable expertise in computer networking and time sharing, computer security, and other areas is documented in [17].

[13] The computer firms that attended these meetings were IBM, Sperry Univac, Control Data, Burroughs, and RCA.

representatives, future CDC ALS project manager Fred Laccabue, remembers discussions at Hanscomb Field far differently.[14] He and other CDC officials believed the plan had a major design error: that the Air Force was writing the central control system using COBOL rather than machine code. COBOL could not approach the efficiency of assembly language (native language of a particular computer). The use of COBOL was to try to provide operating system-like functionality, such as job scheduling, to ALS. The small CDC team, including Laccabue, expressed a few general concerns and requested a subsequent private meeting with the developers of the RFP in 1969 [18]. There, the CDC team explained in detail what they saw as problematic elements (including the use of COBOL) in the RFP specifications that would be "real impediments to achieving success for the program." [18] In Laccabue's words, the CDC team was "not so politely rebuffed." [18] They were told that the Air Force had "employed many experts in the computer field" regarding the central control system and unified database, and were "absolutely confident they were going down the correct path." [18]

Elliott correctly acknowledges that CDC had emerged as the number two computer firm (behind only IBM) in profitability by the late 1960s, but he generally presents CDC in an unfavorable light. He suggests that CDC was a relatively new, untested, and risky company that "almost went bankrupt before its first computer was delivered." [2] This vision of CDC contrasts sharply with what CDC was when it bid for ALS. By the late 1960s, CDC had emerged as the world leader in supercomputers, owing in large part to the skill of arguably the most gifted computer design engineer alive, Seymour Cray. It had established a profitable computer peripherals division and was thriving as it concomitantly sought to extend its success in scientific computing and build its capabilities in business data processing hardware, software, and services [19]. Though Elliott mentions a qualifying ALS benchmark test for which specifications were distributed in July 1969, he fails to disclose that CDC was the only firm to pass this ALS test in its first, and originally its only planned, incarnation. IBM (using dual System 360/67 computers), Sperry Univac, RCA, and Burroughs all failed [2]. Despite the Air Force's overarching goal to move quickly with ALS (after all, the nation was at war in Vietnam), it decided to delay the contract award process roughly six months to allow other computer firms to qualify to bid. This extension upset leaders at CDC [18]. Control Data had invested heavily to set up a huge complex at their Sunnyvale, California facility to house the mass storage disk drives and mainframe computers to meet the benchmark. In fact, they had computers and mass storage sufficient to run the test transactions in roughly twelve minutes, when the benchmark was sixteen minutes [18]. The dozen or so Air Force officials had already watched the other firms fail, and initially, were elated to see CDC's results. Top Control Data officials believed extending the testing for a half year for a second round was just a means to drive down the end price for the contract [18].

Ultimately, IBM dropped out of the competition due to conflicts regarding liability specifications, and Sperry Univac (having passed on its second try) was the only firm

---

[14] Evidence indicates not only CDC, but other potential vendors cautioned "that there was a serious question on the availability of adequate software and that it might be beyond the state-of-the-art." [6]

competing against CDC for ALS [20].[15] CDC was awarded the initial computer and software contract for $87.4 million in April 1972 [21]. This was one of the very largest government computer contracts to that time, but well justified to AFLC leaders—they believed ALS would result in savings of $250 million, through automation and greater efficiencies, in the succeeding decade [22]. It was also, by far, CDC's largest contract to date [23].

The Control Data contract called for 21 CDC Cyber 70 computers, three each at the AMAs, with the remaining six at the Logistics Command headquarters at Wright-Patterson. The agreement specified a complex monthly lease with right-to-purchase structure for the stipulated seven-year life of ALS [21]. Approximately 700 AFLC personnel were working on ALS by this time at Wright-Patterson and about 400 at other AMAs, and these numbers escalated rapidly in succeeding years. The Air Force would develop the central control system and applications software, while CDC would contribute a transaction-based operating system named ZODIAC. The central control system, written in COBOL, was an essential piece of *system* software standing between ZODIAC and any applications software [24].

The CDC Special Systems Division ALS team, managed by Laccabue, was overseen by a high level ALS Program Management Board at CDC that included Robert Price, President, Special Systems and Services Division, and other senior CDC executives [25]. Laccabue joined CDC as a programmer analyst in 1960 after a short stint as a dynamics engineer at Convair. He had participated in the preparation, bidding, and early phases of a CDC Special Systems Division project for the Worldwide Military Command and Control System. This involved discussion and preliminary research on transaction oriented operating systems. His experience, on the promising but young technology of transaction oriented operating systems, made Laccabue a strong choice for CDC's ALS project manager position [18].

Given CDC leaders' uneasiness with the central control system software being programmed in COBOL and with some other Air Force specifications, they carefully crafted their proposal and the lease-to-purchase contractual terms. Purchase credits were greater if the Air Force executed a purchase at an earlier date. In effect, if the Air Force delayed in purchasing the CDC Cyber 70 computers (beyond the initial target date), it would end up paying significantly more money for the systems. Control Data contract vice president H. D. Clover was the mastermind behind this structure that protected CDC if the AFLC could not deploy the system successfully, and on schedule. As Laccabue later recalled, the GAO subsequently prohibited government contracts being structured this way. Laccabue emphasized that this was not to pull one over or be unfair to the Air Force, which CDC hoped to do future business with; it was merely a protective mechanism given what Control Data leaders saw as a risky project and CDC's dependence on the Air Force for project success [18].

Unlike the initial harsh response CDC received at Hanscomb field at the pre-contract phase, some top AFLC officials subsequently understood CDC's concerns. However, they thought these challenges could be readily overcome by AFLC data processing personnel. The Air Force had thousands of programmers and initially the focus was on central processing unit (CPU) cycle time: beating benchmarks. The Air Force was far from alone in the late 1960s and early 1970s in underestimating the

---

[15] RCA and Burroughs did not attempt second round benchmark testing.

delays and challenges of major programming projects. Project delays and cost overruns were essentially the norm for this time period—it was a lesson that most organizations learned the hard way, and a fundamental element to a perceived and real "software crisis" throughout much of the 1960s and 1970s. When IBM's OS/360 project manager Frederick Brooks documented challenges from major programming projects—including the lesson that adding additional programmers to a late project can actually further delay a project's completion—in the *Mythical Man Month*, the book became an instant classic [26]. Brooks had outlined fundamental pitfalls in managing software engineering that all too often victimized organizations. The book, which changed the landscape of software engineering, drew on lessons from the 1960s, but was not published until the mid-1970s.

In 1972, just after awarding CDC the contract, AFLC officials visited CDC's Sunnyvale facility. Laccabue emphasized the substantial work yet to be done before deployment. AFLC officials were surprised, but generally understood and took the attitude that with substantial cooperative effort, these challenges could be met. Several weeks later Laccabue met with General James Bailey, who in July 1971 had become the deputy chief of staff, comptroller, Headquarters AFLC, overseeing ALS for the Air Force [27]. He conveyed to Bailey that to be successful, the central control system had to be written in machine code. He offered to submit a CDC proposal to add this. General Bailey's view was that CDC and the AFLC continue to work closely together, and as they got to next steps, the AFLC would entertain providing a contract to CDC to write the central control system in machine code—but this never happened. While CDC soon began to deliver Cyber 70 computers to Wright-Patterson and the other AMAs (in the second half of 1972), the AFLC leaders, in spite of CDC efforts to inform them, did not grasp the gravity of the software problem. Late in 1972, in an early test at Kelly Air Force Base, Texas, a simple transaction to replenish a part got caught in a loop and took 23 hours. This led to AFLC leaders' general recognition of the severity of the problems with ALS. The system clearly did not work efficiently in real-time. The AFLC leaders made the decision to redefine the project.

This redefinition of the ALS project occurred during renegotiations held between the AFLC and CDC in January and February of 1973. Terms included giving CDC later delivery dates in exchange for added work on the software and hardware [6]. While Elliott generally presents CDC as falling down on original specifications, this is, at best, distorting. There were some issues with hardware reliability, but the greatest problems were with the software, and system integration—namely integrating the Air Force's central control system and CDC's ZODIAC. These problems extended directly from what the CDC team had expressed concerns about to the Air Force from the bidding process forward. At the renegotiation, the AFLC abandoned the concept of a unified data base and real-time computing system in favor of multiple databases and batch processing. Both CDC's and AFLC's work on the project was significantly redefined to build a workable batch processing system. CDC had little choice but to agree to the change and receive the extensions—but it was a bitter pill, as CDC had delivered a system specifically designed to handle real-time processing efficiently that would now be used for batch processing 90 percent of the time [28]. The redefinition greatly reduced the sophistication of the ALS system [7]. Both Elliott's study, and a GAO assessment report after ALS was shut down, stressed that CDC got the better of the Air Force in this renegotiation [2][6]. CDC would provide a version of its SCOPE operating system (developed for its 3000 and 6000 series of computers)[6] and "Multiple Data Base" (originally developed by CDC for the NASA Skylab project) that were better suited to a batch environment than ZODIAC (a transaction-oriented processing system)[24]. CDC leaders saw its offer of these two programs as helping with the Air Force's ALS "get well plan."[24]

General Bailey retired early in 1974 and General Louis Alder succeeded him as deputy chief of staff, comptroller, Headquarters AFLC, overseeing ALS beginning in May 1974 [29]. By August, with continuing ALS problems, Adler presented a briefing to the AFLC Commander, General Catton. Catton proceeded to halt future programming efforts on ALS on August 23, 1974 and launch an internal assessment [2]. He wrote to General David Jones, chief of staff, Air Force that he had "placed too much confidence in General Bailey," who had long been "overly optimistic" about ALS [2]. Catton summarized three reasons for the projects severe shortcomings: 1) the unified data bank was unproven (size precluded efficient processing); 2) the hardware and the software (both vendor and Air Force) were generally untested; 3) concurrent development of the operating software by the Air Force and CDC has "proved impractical." [2] Catton retired later that year and cited the ALS implementation effort as his greatest career disappointment [2]. The problems were soon disclosed to the US Congress and the Secretary of Defense [2].[16] A congressional investigation followed and in December 1975 the ALS project was terminated. From planning to shut down, it had been a nine-year effort with an expenditure of approximately $250 million. The Secretary of Defense, Donald Rumsfeld, soon authorized a plan for a new inventory management system using machine independent software [6].[17]

Once the whistle was blown on ALS, the GAO completed a report on the project [6]. While the report does cite some issues with CDC hardware and software, it overwhelmingly places the blame for the project's failure with the Air Force. In summary, the GAO stated:

> *Many factors contributed to the Air Force's unsuccessful system design and development efforts. But the major factor was that the Air Force did not manage the system as a complex, high-risk program that stressed computer equipment capabilities and software technology. Although the Air Force was aware of potential technological... [problems]... it did not exercise prudent management when system development problems occurred [6].*

Even though the overwhelming blame in the GAO Report was placed with the Air Force, CDC leaders took issue with the fact that the report neglected to emphasize (mentioning only briefly) the "serious inadequacies" of the original Air Force specifications. In a response letter to the GAO, Robert Price cited a government contract panel that had evaluated and reported on ALS in 1974 stating:

---

[16] An earlier disclosure to Congress was made in April 1974 by two employees at McClellen Air Force Base (Sacramento, California) in a letter to California Senator Alan Cranston. This correspondence indicated that the project was a waste of money and not achieving its goals.

[17] Subsequent efforts in applying advanced IT to logistics moved at a more measured pace.

The Contractor [CDC] has delivered the hardware and software in accordance with the contract specifications.... however, there are serious deficiencies in these Government drafted specifications which cause them to fall short of satisfying the ALS objectives [30].

3 Conclusion

The materiality of commands, both the Air Force officers' commands to primary computer contractor Control Data regarding strict adherence to the original design specifications and system limitations on integrating software commands and functionality, were fundamental to ALS' failings. From early meetings with potential bidders, throughout the contracting phase, and into the project, CDC personnel were straightforward about their perception of the limitations and risks involved with specifications established by the AFLC. Given the early frustrations of the CDC team, after being rebuked when they questioned the Air Force computer/software specifications at the pre-proposal stage, CDC managers were particularly careful to document developments and keep AFLC personnel informed about activities and problems with the project. Additionally, both prior to and in the early phases of ALS, RAND advisors had cautioned the Air Force. Ultimately, the combination of the project's complexity and the challenges it posed to the existing state-of-the-art in computing and software, coupled with the well intentioned, but poorly conceived original specifications stubbornly adhered to by the AFLC leaders, resulted in severe shortcomings of the ALS project. Attempts for mid project redefinitions and recovery were made earnestly by the Air Force and CDC, but fell short, and led to the project's demise. The AFLC was left with far more advanced computing infrastructure, but the ALS project was not a cost-effective means to achieve this modernization of computing equipment, and it would be years before the Air Force had a fully operable system roughly comparable to the lofty ambitions initially designated for ALS.

More broadly, the ALS story is particularly meaningful to the history and historiography of computing. It was an unprecedented IT effort in logistics and it was a failed project. Logistics is one of the fundamentally important applications of IT but has received very little attention in the existing historical literature. In James Cortada's trilogy of books surveying the history of computer applications to various industries, *The Digital Hand*, logistics and electronic data interchange (EDI) between organizations make brief appearances (there is significant discussion of logistics and EDI in roughly a dozen pages of this more than 1400- page study) [31][32][5]. In most histories of computing, logistics is not even mentioned. Nevertheless, computers and computer networking were absolutely fundamental to realizing possibilities for efficiencies in logistics, most notably with realizing just-in-time (JIT) inventory management. While contemporary debates exist as to whether IT (including the core area of IT applications to logistics) can still be a source of competitive advantage, it unquestionably was for some organizations (such as Dell and Walmart) in earlier decades [33]. In the 1960s the AFLC was not only the largest purchaser in the US government, but of all organizations worldwide.[18] Given this, it is

---

[18] Furthermore, the AFLC managed Air Force personnel as inventory.

not surprising that the AFLC sought to be a first-mover in applying a massive unified database, and a network of computers operating in real-time, to centralize the task. Like many other organizations of the period, it learned the hard way the great difficulty of managing massive computer programming and system integration projects.

This paper also stands as an all too rare study of technological failure, combating the "progress talk" that frequently dominates histories of technology—a problem articulated by historian John Staudenmaier more than two decades ago [34]. "Progress talk," or presenting technology's history as continual and unwavering progress, holds an even greater stronghold in the history of computing where hardware progress has been quantified as if it were a scientific law (Moore's law—chip capacity/processing capability doubling every year or 18 months) and software, growing to fill ever cheaper memory and add functionality, is frequently perceived as following suit.[19] ALS is a case where project design extended too far out on the technological frontier, and software (with the central control system in COBOL rather than machine code) presented limitations on system integration that could not readily be overcome.

Failures in oversize, massively complex, networked computing systems, such as the FBI's abandonment of a $170 million dollar system in early 2005, continue to persist. At the time of the September 11, 2001 terrorist attacks, FBI computing and networking were horribly antiquated. Using a mainframe operating system three decades old, some field offices were without network connections and unable to transmit digital images of terrorist suspects. A major computer system project was quickly launched to create a networked system, Virtual Case File, to replace the FBI's paper files and aid with tracking criminal cases [35]. Primary contractor, computer services firm Science Applications International Corporation (SAIC), wrote more than 730,000 lines of code and received more than $100 million as the project grew increasingly complicated. SAIC continued to "meet the bureau's requests despite clear signs that the FBI's approach to the project was badly flawed," according to individuals involved with the project and those who later reviewed it for the government [35]. With serious security problems and high error rates, the FBI shut the project down. In a response eerily similar to statements by General Catton and Donald Rumsfeld's in the mid-1970s in aftermath of the ALS failure, FBI Director Robert Mueller took responsibility for "not having put appropriate persons in a position to review... [the]... contract and assure that it was on track."[36] Further, he indicated that "the FBI would now start from scratch, and look for a more updated, flexible system using off-the-shelf software."[36] Similar difficulties have occurred with recent large-scale networked computer systems for air-traffic control, electric energy grid management, and other major IT projects both within and outside of government. These developments clearly suggest that lessons from failures with the design, development, implementation, and oversight of massive computer networked management systems are yet to be fully learned.[20]

---

[19] Moore's Law, initially a speculative prediction by Gordon Moore in 1965, evolved into a company (Intel) and industry-wide (trade association) benchmark. As such, it became a managerial tool shaping investments and outcomes.

[20] One potential lesson is to forego new, path breaking, massively complex, custom systems in favor of existing state-of-the-art systems with proven performance.

References

1. World's Largest Purchasing Agency. Business Machines 10, 62–63 (October 1960)
2. Elliott, W.: To the Third Generation: A History of the Advanced Logistics System, 1967-1977. In: Office of History, Air Force Materiel Command, Dayton, Ohio (1997)
3. Air Force Materiel Command History,
   http://www.afmc.af.mil/library/history.asp
4. GAO Report to the Congress. Potential Problems in Developing the Air Force's Advanced Logistics System. United States General Accounting Office, Washington, DC (1971)
5. Cortada, J.W.: The Digital Hand. In: How Computers Changed the Work of American Public Sector Industries, vol. 3. Oxford University Press, New York (2008)
6. GAO Report to the Congress. Problems with Developing the Advanced Logistics System. United States General Accounting Office, Washington, DC (December 1975)
7. GAO Draft Report to the Congress. Should the Air Force Continue to Develop Its Advanced Logistic System (April 1975)
8. Baum, C.: The System Builders: The Story of SDC. System Development Corporation, Santa Monica (1981)
9. Copeland, D.G., Mason, R.O., McKenney, J.: Sabre: The Development of Information-Based Competence and Execution of Information-Based Competition. IEEE Annals of the History of Computing 17(3), 30–57 (1995)
10. Head, R.V.: Getting Sabre Off the Ground. IEEE Annals of the History of Computing 24(4), 32–39 (2002)
11. Campbell-Kelly, M., Aspray, W.: Computer: A History of the Information Machine. BasicBooks, New York (1996)
12. Campbell-Kelly, M.: From Airline Reservations to Sonic the Hedgehog: A History of the Software Industry. MIT Press, Cambridge (2003)
13. Jardini, D.: Into the Wild Blue Yonder: The RAND Corporation's Diversification Into Social Welfare Research, 1946-1968. Dissertation: Carnegie Mellon University (1996)
14. Abella, A.: Soldiers of Reason: The RAND Corporation and the Rise of the American. Empire. Harcourt, Inc., Orlando (2008)
15. Internal correspondence, RAND Corporation (Willis Ware is likely the author). Memo entitled Inventory Management. RAND Corporate archives (December 1971)
16. Ware, W.H.: Oral History. Charles Babbage Institute, University of Minnesota. Conducted by J.R. Yost. RAND Corporation, Santa Monica, California (August 11, 2003)
17. Ware, W.H.: RAND and the Information Evolution: A History in Essays and Vignettes. RAND Corporation, Santa Monica (2008)
18. Laccabue, F.: Oral History. Charles Babbage Institute, University of Minnesota. Conducted by J.R. Yost. Cupertino, California (December 4, 2009)
19. Worthy, J.C., Norris, W.C.: Portrait of a Maverick. Ballinger Publishing Company, Cambridge (1987)
20. Correspondence (to CDC) from T.L. Meyers (November 9, 1971), Copy provided to author by Fred Laccabue
21. $87.4 Million AF Contract Awarded to Control Data. Electronic News, pp. 1-37 (April 10, 1972)
22. McConnell, A.G. (Lt Colonel USAF): Management of Information Systems Development: The Advanced Logistics System. Unpublished paper completed at the Naval War College, Newport, Rhode Island (May 12, 1975)
23. CDC Newsletter, No. 57 (May 1972), Provided to author by Fred Laccabue
24. Fred Laccabue to G.S. Weller (July 15, 1975), Copy provided to author by Fred Laccabue

25. ALS Project Organization (1974), Document provided to author by Fred Laccabue
26. Brooks Jr., F.P.: The Mythical Man Month: Essays on Software Engineering. Addison-Wesley Publishing Company, Reading (1975)
27. Major General James A. Bailey, http://www.af.mil/information/bios/bio.asp?bioID=4579
28. Letter from Robert Price to Fred Shafer (GAO, Director) (July 18, 1975), Copy of document provided to author by Fred Laccabue
29. Major General Louis O. Adler, http://www.af.mil/information/bios/bio.asp?bioID=4492
30. Letter from Robert Price to Fred Shafer (GAO, Director) (September 10, 1975), Copy of document provided to author by Fred Laccabue
31. Cortada, J.W.: The Digital Hand: How Computers Changed the Work of American Manufacturing, Transportation, and Retail Industries. Oxford University Press, New York (2004)
32. Cortada, J.W.: The Digital Hand: How Computers Changed the Work of American Financial, Telecommunications, Media, and Entertainment Industries, vol. 2. Oxford University Press, New York (2006)
33. Carr, N.G.: Does IT Matter? Information Technology and the Corrosion of Competitive Advantage. Harvard Business School Press, Boston (2004)
34. Staudenmaier, J.M.: Technology's Storytellers: Reweaving the Human Fabric. MIT Press, Cambridge (1985)
35. Eggen, D., Witte, G.: The FBI's Upgrade That Wasn't. The Washington Post (August 18, 2006)
36. Borger, J.: FBI Chief Admits $170 Million Computer Failure. The Guardian (March 10, 2005)

Purpose-Built Educational Computers in the 1980s: The Australian Experience

Arthur Tatnall[1] and Ralph Leonard[2]

[1] Graduate School of Business, Victoria University, Australia
Arthur.Tatnall@vu.edu.au
[2] Department of Further Education, Employment,
Science and Technology – Government of South Australia
Leonard.Ralph@saugov.sa.gov.au

Abstract. The first microcomputers were developed in the late 1970s and soon a wide variety of these machines were available for school and home use. This presented both a marvellous opportunity to improve school education and a significant problem for education authorities in how to provide support for the range of available computers. Several countries, including Australia, attempted to solve this problem by designing and building their own educational computer systems. This paper briefly describes how New Zealand, the UK and Canada designed and built computers for use in schools, and looks in more detail at how Australia started down this path and designed, but did not ultimately proceed to build an educational computer.

Keywords: History of educational computing, purpose-built school computers, Poly, Icon, Acorn, Microbee, Australian Commonwealth Schools Commission, National Computer Education Program.

1 Introduction

The widespread use of computers in schools is now commonplace, but this has only occurred in comparatively recent times, beginning in the late 1970s and early 1980s. Before this time a few Australian schools had some access to a mini-computer or used punch or mark-sense cards at a local university, but these schools were few in number. In an exception to this the Angle Park Computing Centre (APCC) in South Australia and the Elizabeth Computer Centre in Tasmania offered shared computing facilities to all schools in their respective states. The advent of relatively low cost microcomputers such as the Apple][and Tandy TRS-80 in the late 1970 marked the beginning of the growth of computers in schools. These early computers typically stored their software on audio cassettes as disk drives were not readily available and quite expensive until some years later.

An early problem was the diversity of available types of microcomputer, compounded by each Australian state controlling its own school education system. This meant that co-operation between the states was not to be taken for granted. One problem with using these early microcomputers in schools was that while you could show the students what a computer was, and even look at the electronics inside, you could

A. Tatnall (Ed.): HC 2010, IFIP AICT 325, pp. 101–111, 2010.

not do much with them apart from programming and playing computer games as there was not much suitable software available for use in the school classroom. The states of South Australia, Tasmania and Western Australia did have support for development of software and set up the TASAWA consortium agreed to exchange software between these states. This is, perhaps, the first example of organised multi-state collaboration in computer education and the for-runner for the national program, but unfortunately the other Australian states were not parties to this consortium.

Another issue was cultural as what software there was often had an American outlook. An example of this was the simulation game 'Lemonade', available for the Apple][and based on making and selling lemonade from a street stall. While this had some merit in terms of teaching students about one aspect of doing business, lemonade stands are almost unknown in Australia. Another slightly later example is the 'Trash Can' on the Apple Macintosh. In Australia we use a 'Rubbish Bin'.

In the early 1980s the number of microcomputers on the market skyrocketed and education authorities started to see a potential infrastructure problem in servicing the schools that purchased these machines. All this presented both an educational need and a business opportunity and several countries decided to design and build their own school computers. They saw a solution to the educational need in writing their own educational software for these computers, and the business opportunity in having the new computers designed and built locally.

2   Purpose-Built School Computers

2.1   Poly Computer (New Zealand)

Probably the first microcomputer specifically designed for educational use was the Poly from New Zealand. The Poly was designed by Neil Scott and Paul Bryant at Wellington Polytechnic (hence its name) in 1980 as a teaching machine intended for computer assisted learning [1, 2]. Scott and Bryant had recognised a niche market in the education centre and proceeded to exploit it [2] along with a team of engineers and technicians.

Poly-1 was a networkable machine based on the 6809 processor and came with 64k bytes of RAM [1]. The New Zealand government's Development Finance Corporation partnered with Progeni Computers [3] to form Polycorp. Poly was manufactured by Polycorp New Zealand, and became available in 1981. Polycorp had worked towards getting assistance from the New Zealand government, but this fell through. Polycorp worked with a number of New Zealand teachers to produce and refine courseware for a variety of teaching areas. The main problem with Poly was its cost of around $8000 (NZ) which was considerably higher than competitors such as the Apple][. Smythe [2] and Harpham [3] claim that the Poly computer was eighteen months ahead of the Acorn BBC Micro computer and with government support could have become highly significant on the world scene.

2.2   Acorn BBC Computer (UK)

The BBC Micro was designed and built by Acorn Computers in the early 1980s for use in the British Broadcasting Corporation's *Computer Literacy Project* [4]. The

BBC had noted that with the availability of a growing number of powerful and increasingly less expensive microcomputers on the market, that it would soon be feasible for many people to purchase their own computer at an affordable price [5], and decided to start a computer literacy television series. The BBC needed a microcomputer capable of performing tasks which could be demonstrated in their TV series 'The Computer Programme'. These included: programming, graphics, sound and music, Teletext, controlling external hardware and artificial intelligence [4]. After discussions with several British computer companies, Acorn won the contract to provide a computer for this program and the Acorn Proton (successor of the Acorn Atom) became the Acorn BBC model A. The Acorn BBC model B followed in 1982 [5], based on a 6502 processor and with 32k bytes of RAM.

2.3 ICON Computer (Canada)

In 1981 the Ontario Minister of Education announced a need for computer literacy for all students and set up an Advisory Committee on Computers in Education that would, amongst other things, would draw up plans for an educational computer that would become the standard in Ontario schools [6]. A series of working sessions by various government departments and professional associations during 1981 produced a set of specifications for an educational computer. These included high resolution colour graphics and sound synthesis capabilities (which were only just possible at this time), 64k of RAM and a local area network form of architecture. By 1983 CEMCorp (later to join with Burroughs) had developed a prototype ICON computer to meet the Ministry's specifications. The ICON was quickly nicknamed the 'Bionic Beaver' and the first of these were installed in a few Ontario schools in 1984. The ICON system was designed around the 80186 microprocessor, based on a file server / workstation model with no local storage on the workstations [7] which were housed in a single box that included a keyboard and trackball. The operating system was Unix-like. The Ontario Ministry of Education sponsored the production of educational software and subsidised schools in purchasing their ICON computers from 1984.

3 Designing the Australian Educational Computer

In 1983, the Australian Government's Commonwealth Schools Commission set up the 'National Advisory Committee on Computers in Schools' (NACCS) to plan a National Computer Education Program. The terms of reference of this committee were to provide advice on professional development, curriculum development, software/courseware, hardware, evaluation, and support services [8]. In February 1984 the Commonwealth Minister for Education and Youth Affairs announced an $18.7 million 3 year 'Computer Education Program' that would approach computer education in terms of a broad educational program, rather than simply as an exercise in hardware provision. Nevertheless computer hardware was an important consideration. The Committee believed that Schools Commission funds should be used for the purchase of computer hardware by schools, but as a substantial level of standardisation of equipment was necessary to achieve a balanced and effective National Computer Education Program this should be subject to strict guidelines. In the short term they recommended that Commonwealth funds be provided to support the purchase by schools of *only* BBC Acorn, Microbee 64, and Apple //e computers. This recommendation was later softened so that these funds could be used to purchase a computer on the 'recommended list' drawn up by any given state [9]. In the longer term however the need was seen for Australia to develop an educational computer system of its own.

> *"To meet the long term requirements of schools computing activities in Australia, it is considered essential to embark on a national research and development project that will ensure that appropriate computer systems are available. This ... will involve:*
>
> - *the research and preparation of a set of Educational User Requirements. This is a statement of agreed educational needs to be met by the computer systems;*
> - *the development of a set of Educational Technical Requirements based on the Educational User Requirements. This is a statement of the function, main features and performance required by the user for a system which can reasonably be expected to be available to satisfy the requirements in the planned time period;*
> - *a System Concept Study which involves research and analysis of all practical alternatives to satisfy the Educational Technical Requirements. It includes consideration of development and production options and use of existing items either as they are or in modified form;*
> - *if no existing items satisfy the Educational Technical Requirements, then a development proposal leading to the design and development of appropriate systems is required.* [10 : 44]

The process by which this might be achieved is described in Appendix E of this report [10 :69-72].

> *"The National Advisory Committee on Computers in Schools has recommended that the research, design and development process described below be adopted. This process would involve:*
>
> - *Commonwealth Schools Commission co-ordination and funding of the research and development of educational requirement documentation and;*
> - *Department of Science and Technology co-ordination and funding of computer equipment research, design and development.*
>
> *Elements of the process are:*
> *(a) Commonwealth Schools Commission leadership in the research of requirement document namely:*
> *(1)   An Educational User Requirement; and*
> *(2)   An Educational Technical Requirement.*
> *(b) Department of Science and Technology leadership in equipment research, design and development through:*
> *(1)   A Systems Concept Study; and*
> *(2)   An Australian Design Specification including the design and manufacture of pilot and prototype systems if necessary.*

There were two principal reasons for wanting to develop an Australian Educational Computer: so that Australian school children would have access to suitable, well designed equipment; and to provide a development and manufacturing opportunity for Australian industry [11].

3.1   Educational User Requirements

An **Educational User Requirement Working Party** was appointed early in 1985, and soon provided an interim report outlining the many and varied educational needs of computer users in schools. The report began by considering educational assumptions underlying learning situations in primary and secondary schools, based on a statement from the Schools Commission report:

> *"The emphasis in efforts to integrate information technology in the curriculum should be placed on developing inquiry and problem-solving skills so that students can gain an understanding of the concepts, symbolic terms and language involved. In this way information technology will not be seen as applicable exclusively to any one curriculum area, but as a tool for establishing meaning and communication, for classifying and ordering data and experiences and for opening up new approaches to learning" [10 :25]*

They then listed learning situations in which computer use was considered appropriate, including: brainstorming, inquiry learning processes, 'dialectic' problem-solving, 'procedural/technical' problem-solving and process writing. They then went on to consider scenarios of activities and their organisation, including: classroom interest centres – a primary school scenario; co-operative large group use; flexible and varied modes of classroom use; project group use; gathering, organising and analysing information; developing language skills; computer assisted learning; expert systems; using computers as a tool in existing subject areas; studying computer science; using computers in special education; whole school use of computers; and evaluation of learning [12]. The report then attempted to draw *user requirements* from each of these. For instance in the case of co-operative large group use:

> *"User requirements as a consequence of this large group of learners reacting to a single monitor would include an emphasis on the need for a large clear video display visible to all students in the group and the use of colour, graphics and sound/music capabilities."* [12 :12]

In summary the report highlighted: the need for a common user interface, the need to consider a variety of user environments, the need for a modular compatible construction so that hardware and software can be added and subtracted later as required and a need for adoption of current recognised standards [12].

3.2   Educational Technical Requirements

Education in Australia is the responsibility of the State Governments, the Commonwealth's main role being in the co-ordination and funding of special projects. The **Technical Requirement Working Party** [13] was set up in 1985 as an 'expert' committee with membership reflecting the range of relevant groups and interests: David Woodrow (St Peter Luther College, Queensland), David Ashmore (Director, Information Technology, Department of Industry, Technology and Commerce), David Nicholls (Assistant Director, Information Technology, Department of Industry, Technology and Commerce), Paul Jenner (Senior Education Officer, Computer Education Unit, NSW Department of Education), Les Keedy (Newcastle University), Ralph Leonard (Co-ordinator of Computer Resources, Angle Park Computing Centre, South Australia), Jim Park (Head of Data Switching Networks, Telecom Australia Research Labs), Andy Quaine (Computer Science Department, Australian Defence Force Academy), Jim Sully (Superintendent of School Computing, Education Department of Western Australia, Arthur Tatnall (Educational Computer Systems Analyst, State Computer Education Centre, Ministry of Education, Victoria) and Steve Murray (Chief Education Officer, Computer Education Program, Commonwealth Schools Commission). The Committee met for a total of 18 days in the period June 1985 to March 1986, finally publishing its report in mid-1986.

3.3 Recommendations

The report has two main sections, one detailing the technical requirements and the other suggesting possible implementations [13]. In the requirements section the Committee endeavoured to keep things as general as possible and not to mention specific figures, such as 64k RAM, unless this was unavoidable. After almost 25 years this section still looks remarkably up to date.

It was considered that an implementation of these requirements would need to satisfy at least three types of use: personal, classroom and school-wide. The implementation guide suggested that these could be catered for by a family of compatible systems, having a common user interface [13], and that at some stage in the future the way should be left open to connect these systems to computing facilities at the district, regional, state or national levels.

3.3.1 Personal Systems

The system intended for individual use should be totally portable so that it could be used by students in a classroom at school, in the school grounds, at home, on the bus when travelling, or anywhere else required. It was considered likely that use by an individual student for word processing would be its major applications, but that it would also be used to perform applications such as use of spreadsheets, educational simulations and the manipulation of small databases. The personal system would need to be totally upward compatible with classroom and school systems [8, 11].

3.3.2 Classroom Systems

The computer systems normally used in the classroom need not be portable, but should still be able to be moved around within the school. They should be able to be configured to perform a much wider range of tasks that the Personal System, including all those currently asked of school computers. They should be easily expandable, possibly with plug-in cards or connection of external expansion units.

MODULES

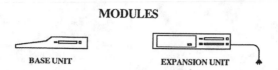

Fig. 1. Possible implementation using a Base Unit and an Expansion Unit

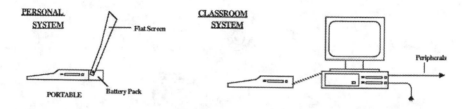

Fig. 2. Implementation linking a Personal System into a Classroom System

In one possible implementation, a Classroom System could be built up by adding appropriate extra components to a Personal System. The provision of normal (non-portable) colour monitors to which a Personal System could be connected would be one variant of this implementation.

Another implementation may include the attachment of an external expansion unit. Although not the only way this system could be implemented, the use of a common base unit which would also be the Personal System was one option [8, 11].

3.3.3 School Systems

School Systems would comprise a network to enable Personal and Classroom Systems to be connected to each other, and to devices such as printers, mass storage devices, special purpose peripherals, and remote computers [13]. A School System was envisaged to be a transparent system with a number of connection points in each classroom and around the school [8, 11] so that students could plug Personal Systems into connection points to use a printer or to up or down-load software or perhaps an assignment. A number of Classroom Systems could be connected to the School System to facilitate use of software, sharing of resources, and the sharing of common data. At any time the School System could be decomposed into its individual modules to form a number of Classroom and Personal Systems.

An educational computing scene envisaged for the future was one where each student would own a Personal System. These would be built on contract for the government and purchased, perhaps on a long term leasing basis, by individual students. Classroom and School systems would be purchased by schools using government funds. Students having constant access to a Personal Computer System would revolutionise the education system and make many of the dreams of computer educators possible.

SOLUTION (C)

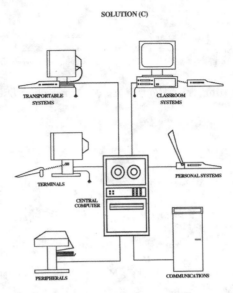

Fig. 3. One possible implementation of a School System

3.4   Building the Recommended Educational Computer

There were two principal reasons for wanting to develop an Australian Educational Computer: so that Australian school children would have access to well designed equipment and to provide a development and manufacturing opportunity for Australian industry. It was generally supposed that an Australian company such as Microbee, which produced a CP/M computer used in many Australian schools, would be a likely manufacturer. The next step in the process should have been the setting up of a System Concept Study to be followed by a Development Proposal, but at this stage the project ran out of steam, as the 3 year Government funding for the program was at an end and further funds were not made available. The System Concept Study and Development Proposal were thus not taken any further [8].

4   Reflections on a School Computer That Was Never Built

One of the strengths in the work undertaken to design Australian School Computer Systems was that the work of the Technical Requirement Working Party lay within a program that intended to be far-reaching. The News Release by the Minister for Education and Youth Affairs, announcing the $18.7 million Computer Education Program on 19th February 1984, stated:

> *"We are going to approach computer education in terms of a broad educational program, rather than simply as an exercise in hardware provision. The central themes here are building a capacity to generate Australian content with sound educational values across a broad range*

of school subjects; and developing a teacher force capable of using computers to the advantage of all children in school."

Hence the design was driven by strong principles of usability in a wide range of educational contexts. The Technical Requirement Working Party was given instruction to be particularly forward thinking and

- *"be guided by the user requirements and not restricted by available or predicted equipment, nor be unduly influenced by the current state of the art;*
- *take into account the relevant documents from the Australian States and overseas*
- *provide some indication of the relative priority that is attached to various technical characteristics, at the least indicating those which are essential, and those which are desirable but not essential."* [13 :1]

These instructions were matched by the selection of members of the Technical Requirement Working Party which was assembled with *"detailed experience of the requirements of Australian states, detailed up-to-date technical expertise, and a good understanding of the future directions of computers in education"*. [13 :1]

The Working Party's recommendations, described in section 3.3 above, illustrate a successful outcome in looking beyond the existing state of the market during the project and separating the desirable elements of a school computer system into a set of modular components.

However, even as the Working Party was finalising its report, new entrants to the personal computer market were extending the state of the art and rapidly progressing beyond the recommendations. The Amiga computer was released in the latter half of 1985 and took the expectation of colour displays and graphic capability beyond what the Working Party had envisaged. Atari Corporation also produced a much enhanced Atari ST computer at about this time. Nevertheless, after a quarter of a century, many of the recommendations remain consistent with the current practices regarding choices of equipment for educational environments.

What might have happened if the project had continued to the planned consequent stages of Development Proposals, Australian Design Specification and finally procurement? That remains an intriguing "what if" question. Not long after the release of the recommendations in June 1986, the three year funding for the Commonwealth Schools Commission's Computer Education Program was up and the program was discontinued. After the funding made available in the triennium from 1984-1986 was used, no further funding was allocated in the following years. Had further funds been available it is likely that the project to design and built an Australian Educational Computer would have continued. But in the longer term, would this have been a good thing?

A problem faced in Canada with the Icon computer was that after the government had spent so much money on one particular educational computer the idea of looking at other alternatives was not an appealing one. In a sense, no matter how good it had been at the time, Canada was stuck until the early 1990s with what it had designed and built in the early 1980s. This would not have been a problem if the technology had been static or even if it had been evolving slowly, but at this time a number of major changes occurred to the microcomputer market. While in the early 1980s there had been a large number of microcomputers available and potentially useable in schools, by the second half of the decade two significant players had begun to emerge and to displace all the others: the Apple Macintosh and the IBM PC (and compatibles). It was not long after this that these were the only microcomputers to occupy a significant place in school education, particularly after the advent of the Windows operating system (on IBM compatible PCs). An important consequence of this was that pretty much all software development occurred only on one or other, or both, of these platforms.

Another major development was the growing dominance of 'application software' packages including a word processor, spreadsheet, presentation software, graphics package and database manager, exemplified by the increasing significance of Microsoft Office. The problem was that Microsoft Office was available only on the Macintosh and Windows platforms and not for other microcomputers. As it is unlikely that an Australian Educational Computer would have been built to use either of these operating systems, it may quickly have become a Neanderthal that, although worthwhile in its own right, could not evolve further in line with emerging trends. Would this have been the case? What if the Australian project had proceeded right through to manufacture? How would building such a computer have stimulated the Australian computer industry? These are more intriguing 'what if' questions to which we will now never know the answers.

5 Conclusion

From the late 1970s to the mid 1980s several countries attempted to solve the problem of providing useful computer systems for their schools by designing and building their own Educational Computer Systems. This paper has briefly described the Poly from New Zealand, the Acorn BBC from the UK and the ICON computer from Canada. That Australia also started down this path to design and build its own educational computer, but did not complete the exercise, has been the subject of this paper.

In retrospect, was the Australian exercise a waste of time and money? We suggest that it was not a waste of time and money as some useful results emerged from this project. The reports published by the two working groups are of value, even today, as were the interstate connections forged during the process of researching and writing these reports. On the other hand, given the benefits of hindsight, we also suggest that it was probably a good thing that the project stopped after the Educational User Requirement and the Educational Technical Requirement working parties had completed their work and published their reports. It was probably a good thing that it did not continue to the stage of a Systems Concept Study, Australian Design Specification and manufacture of the computer. This, however, we will never know for certain.

References

1. Editorial, Executive's Fighting Pledge. Bits & Bytes, New Zealand, pp. 13-15 (1982)
2. Smythe, M.: The Poly 1 Educational Computer. Kiwi Nuggets Forum 2007 (2007), http://www.creationz.co.nz/kiwinuggets/2007/03/poly-1-educational-computer_07.html (cited 2010)

3. Harpham, P.: Poly and Progeni (2007),
   http://www.mail-archive.com/ada_list@list.waikato.ac.nz/msg00266.html (cited 2010)
4. Wikipedia. BBC Micro (2009), http://en.wikipedia.org/wiki/BBC_Micro (cited 2010)
5. Graça, G.: Acorn BBC. Old Computers.Com (2009),
   http://www.old-computers.com/museum/computer.asp?c=29&st=1 (cited 2010)
6. Goodson, I.F., Mangan, J.M.: The Genealogy of the ICON. In: Goodson, I.F., Mangan, J.M. (eds.) History, Context, and Qualitative Methods in the Study of Education, pp. 207–248. University of Western Ontario, Canada (1992)
7. Wikipedia. Unisys ICON (2009),
   http://en.wikipedia.org/wiki/Unisys_ICON (cited 2010)
8. Tatnall, A.: Designing the Australian Educational Computer. Education 110(4), 453–456 (1990)
9. Tatnall, A., Jenner, P.: How State Education Authorities Recommend Computer Systems for Use in Australian Schools. In: Australian Computer Conference (ACC 1986). Australian Computer Society, Gold Coast (1986)
10. Commonwealth Schools Commission, Teaching, Learning and Computers. Report of the National Advisory Committee on Computers in Schools. Commonwealth Schools Commission, Canberra (1983)
11. Tatnall, A.: The Growth of Educational Computing in Australia. In: Goodson, I.F., Mangan, J.M. (eds.) History, Context, and Qualitative Methods in the Study of Education, pp. 207–248. University of Western Ontario, Canada (1992)
12. Commonwealth Schools Commission, Australian School Computer Systems: Educational User Requirements. Commonwealth Schools Commission, Canberra (1986)
13. Commonwealth Schools Commission, Australian School Computer Systems: Technical Requirements. Commonwealth Schools Commission, Canberra (1986)

And They Were Thinking? Basic, Logo, Personality and Pedagogy

John S. Murnane

The ICT in Education and Research Group
The University of Melbourne, Australia
jmurnane@unimelb.edu.au

Abstract. This paper is concerned with some limited aspects of the history of two programming languages purpose-designed for students learning to program digital computers: Basic and Logo. The focus is the very different educational aims and philosophies of the originators of these languages. They are compared and their early use in schools sketched. While the reasons for teaching students to write programs were initially based on experience in programming digital computers for non-educational use, despite extensive research and publications, it would seem that the teacher of today is not in a much better position to justify teaching programming than the original pioneers.

Keywords: Computer education; introductory programming languages; history of computing.

1 Introduction

> *The introduction of yet another language clearly deserves critical examination.*
> Feurzeig, Papert, Bollm, Grant, and Solomon [1 p12]

This paper is concerned with the history of programming languages purpose-designed for students learning to program digital computers. A proper treatment of this topic would run to a small library and deal with the ideas and intent behind the form of the language, research on its success in educational use, examples of classroom use and modifications made as a result of experience, so stringent selection was necessary. My main interest is in the intent of the creators the languages, so I began with the idea of examining the educational concepts behind the development of several of them, but in the end found space for only two, Basic and Logo, and then in a very constrained form.

The creation of a programming language of any sort is a complex business and the province of a special elite in the world of programming. Yet the difficulty of the creation of a language for fields such as business or mathematics for which there is an existing set of well tried models, pales into insignificance compared to the task of creating one for educational purposes: a space where a choice must first be made between various pedagogies, all with their own built-in advantages and disadvantages, advocates and detractors, before even the form of the language is decided on. Nor will the educational aim be simple. Is it to be a language to introduce the forms and

A. Tatnall (Ed.): HC 2010, IFIP AICT 325, pp. 112–123, 2010.

disciplines of programming itself, or is it to facilitate a more general development in problem solving and analytical and logical thinking? Are we teaching computing and its applications, or are we using programming for some wider educational purpose? Ham [2 p34] believes that, when it comes to the use of IT in education, "even within the educational policy and research communities, people do not necessarily agree on the questions which are worth asking."

I wish to examine the published thoughts of some of the earliest pioneers who dared to swim in this very complex sea.

2   The Promise

> *We have learned how to work with the computer in solving a problem, rather than submitting a problem for machine solution.* Kemeney and Kurtz [3 p22]

Even in the new centaury it is easy to find material critical of the use of computers in schools, from Cuban's 2001 [4] characterisation as being "oversold and underused," to Cox (2010) [5 p16] who found "the actual integrated use of IT by the teachers is much lower than might have been expected from so many sustained national and international programs." But even allowing that not all teachers use IT brilliantly, it is hard understand Munro's [6 p47] criticism that "microcomputers were introduced into educational institutions with no prior research and with no educational rationale for their use." True, they were. When something is new its introduction cannot be based on research, and the educational rationale must be the belief of the teacher in the promise the new idea holds. It was the promise the new world of programming held for all sorts of educational and cognitive advances that attracted the pioneers of educational programming languages. "Programming," declared Ershov in 1981 [7 p1] is "the second literacy."

Looking back at their a-priori positions on the benefits that writing programs could bring to their students, one finds a remarkable unity of spirit. Cynically, it could be argued that in the 1960's, apart from some rather inflexible Computer Assisted Instruction (CAI), and some (mostly non-interactive) simulations, writing a program was about the only educational thing you could do with one. But the pioneers, and particularly those with a hand in writing specialised educational languages, such as Kemeny, Kurtz, Feurzeig and Papert, were all convinced that great educational advantages would come from programming a digital computer, although often for different educational and cognitive reasons. Weyer and Cannara [8 p3] put it this way: "If, by a free interpretation of Church's thesis, any ideas which can be formalised may be studied concretely via a computer program, then, by learning programming in full generality, students can learn to construct laboratories to study any ideas they wish to think about."

Basic and Logo were both written in the 1960's, were specifically designed for educational purposes, and are still in widespread use. A fascinating speculation, now difficult to resolve from material published of the time, is how much the designers were influenced by their educational philosophy, and how much by the educational environment in which they happened to be, and the available tools. Feurzeig and Papert, operating in the Artificial Intelligence Laboratory at Stanford, the home of Lisp, not only had an example of a language congruent with their educational ideas, but one in which their new language could be written. Kemeny and Kurtz had two very small computers from which they hoped to assemble a useful system, neither of which had so much as an operating system.

The decade that produced the first educational computer languages is now 50 years in the past. With the singular exception of anything written by Seymour Papert, looking back through the papers and reports leaves a distinct impression of teachers striving: striving towards goals imperfectly grasped, using computing equipment barely up to the task and hampered by primitive translators, operating systems and input/output devices. (Papert knew what he was doing from the outset.) Yet the *overall* feeling is of high optimism, more positive than one finds across the literature in 2010. The pioneers *knew* that programming a computer had educational benefits, and were going to set about proving it to the world.

3 Basic

Our goal was to provide our user community with friendly access to the computer. Kemeny and Kurtz [9 p534]

Seymour Papert always complained that besides being a poor language, Basic had been left for the academic community "to pick up, like cast-off clothing" [10]. There is some weight to this, for Kemeny and Kurtz, Basic's authors, do concentrate in their publications on Basic as Computer Science and do not say much on its pedagogical or curricular aspects. Their Final Report to the Course Content Improvement Program of the National Science Foundation, who financed it, is titled "The Dartmouth Time Sharing System" [3] rather than something suggesting *educational* advance. It gives the reasons for teaching programming to college and secondary students as:

- The need for more people to learn to program because of the key roles computers play in "'business, industry, government and all forms of research."
- To change the attitude "of the typically intelligent person towards computers," which they characterised as "a mixture of fear and superstitious awe." (One wonders how much has changed!)
- To put "the computer at the fingertips of the Faculty." [3 p1]

There is little here to explain just why writing a program might be educationally advantageous. They simply state "the hard question was not 'whether' but 'how'" [9 p518]. They do give many examples of programs written by students across a range of disciplines, but leave it largely to the reader to decide on the cognitive benefits that accrue. Significantly, they did find that in a programming environment, students were more likely to share ideas [3 p16]. They also characterise computers as "a magnificent means of recreation," (p. 8) something which in 2010 threatens to overshadow their significance for learning.

Kemeny and Kurtz strongly distinguish between using computers for instruction and having students write their own programs. While they saw the possibilities of CAI, and even thought "they ought to do more," they saw much more potential when "the student is the teacher and the machine learns" [3 p11], noting that "by being able to program certain processes, the student necessarily shows a through understanding of the process" (p. 27). Both these themes run through computer education to the present day.

Understandably, Kemeny and Kurtz devote most of their various publications to Basic itself and the ground-breaking time-sharing system shared between two computers that went with it. Written by sophomore Michael Busch and junior John McGeechie [3 p5], this system was the element which made the project economically possible and educationally viable, an example of what can be achieved by enthusiasts too young to know that what they were doing wasn't supposed to be possible.

Basic was an acronym for 'Beginners All-purpose Symbolic Instruction Code.' Dartmouth, where its originators taught, attracted students who were "not generally interested in the Sciences," so it was designed for those studying the liberal arts [9 p518 & 522] "as an extremely simple language that can be quickly mastered by a novice" [3 p3]. They considered that Fortran had "many disadvantages for the novice and occasional user," meaning largely there was too much fussy detail to remember, and "decided that we would improve it" (p3). Considering other languages available or planned, they thought the Algol compound statement introduced too many complexities for beginners [9 p538] and wanted something much smaller and more general then Cobol. As long as the user is content with real numbers (which Kemeny and Kurtz considered removed the need for typed variables), and happy with two-character variable names (forced by the exigencies of the computers), Basic can be said to express mathematical ideas generally as well as Fortran. Indeed, it has the added power of matrix operations, and with the extension to string handling it came much closer to being truly 'general purpose' as well. It has also shown remarkable longevity, widespread use in commercial applications and a capacity to accept extensions gracefully.

Perhaps the best expression of the benefits of teaching students to program comes from Thomas Dwyer [11-14]. Dwyer extended Dartmouth Basic to make it a better language to teach *with*. This may sound contradictory and clash with the educational ideas of Basic's creators, but actually it was designed to reinforce them.

Kemeny and Kurtz discuss the advantages of having a student teach the computer. Papert's *Mindstorms* [15] can be seen as an extended plea for learner control. Dwyer considered that students learn best when they are teaching other *students*, so he wanted his students to write programs for others to learn *from*. Hence his characterisation of educational programming as having two modes: Dual and Solo. Dual mode consisted of using a computer to learn with something programmed by someone else—what today would be covered by courseware and information retrieval and processing. Solo mode was, initially, writing programs for your own use, but then going on to write programs to teach different parts of the curriculum to others. Here he was picking up Kemeny and Kurtz's ideas on needing to understand a process if you are to write a program for it. Dwyer [12 p220] explained this as:

> *a learning situation which develops advanced cognitive and motor skills for students of quite varied backgrounds, and which also involves many affective elements, but which relies heavily on technology for achieving these ends. While it is clear that dual instruction is essential (one does not recommend that a student immediately go out in an airplane and 'do his own thing') it is equally clear that the student will optimise his benefits from the dual mode if*

he knows he is preparing for a solo flight. He knows, in fact, that he can eventually exert more influence on his learning than his instructor ... computer technology in education should invite similar control at all levels. It should, in particular, invite the student to 'go behind the scenes' (possibly acting in concert with teachers) at any time they elect. There should be no secrets, on one-upmanship of the adult world over the student world.

To go behind the scenes and do your own thing you have to be able to change the system. In 1971 it meant you had to be able to program. These days with the advent of Web 2.0 and all sorts of multi-media vehicles it's a bit different, but the principle still holds, more strongly if anything. diSessa [16 p164] advocates a similar hierarchy. Papert considered the "proper use ...of drill-and-practice programs" was something for other students to write [17 p4–1]. It's also interesting that Dwyer emphasised, as his first principle, the essential "social character of human learning," though I doubt that in the USA in the 1960's he'd ever heard of Vygotsky.

Despite its attractions, Dartmouth Basic has always been heavily criticized for its simplicity and the lack of structure and formal rigor of programs written in it. The first definition of Algol, with its ground-breaking ideas and structure was published in 1958, and an Algol compiler was available at Dartmouth by 1967 [3 p8]. It would seem that Kemeny and Kurtz's desire for simplicity overruled thoughts of enhanced Scientific (or Mathematical) precision.

4 Logo

In original conception, Logo was conceived as a form of Lisp suitable for beginners to write programs in. The 'Lo' in the title suggested 'Logic,' and the earliest versions of the Logo system were written in Lisp. It is curious then that early papers [1 p1, 18 p1] make it clear that it was "expressly designed" for the teaching of Mathematics. At that time there was no Turtle Geometry, and indeed no arithmetic functionality beyond addition and subtraction. Brown and Rubinstein [19 p3] flatly describe it as "non-numeric."

Early Logo was very simple. Like Basic, it came with built-in editing and file manipulation commands. If these are ignored, the 1971 version consisted of just 26 'operations,' five of which accessed the calendar and clock, and 15 'commands.' Most operations were concerned with program logic or list manipulation. An essential part of the design was to produce a language of such simplicity that it forced users to write their own library of commonly used routines, such as multiplication and division. From such a library, complex programs could be built. "Ideally, by the end of the course, each student would have created his own extended version of Logo" [19 p10]. If you exclude the rich set of mathematical functions, contemporary Basic was even smaller, but for a different reason: to make the language as easy as possible to learn. (And remember Basic was designed for the 'non-mathematical.')

There is no mention of the degree of difficulty inherent in learning to program generally, and certainly not in learning Logo, in Feurzeig's papers. The entire emphasis in *The Final Report* [1] is on the difficulty of learning *Mathematics*, and how Logo was developed to make that easier. The programming language followed as a result of a specific educational need. The designers of Logo intended it not only as a vehicle to express Mathematical ideas and make Mathematical concepts concrete, they saw it as a meta-language in which to express Mathematical thought [1 p5]. They wanted a "standard, teachable terminology to discuss the *heuristic aspects* of mathematical activity concerned with the art of solving problems" (p. 5, their emphasis). Here is the origin of Papert's often expressed need to teach 'thinking about thinking' [17 p2] and the decision to write a computer language whose primitives and predicates inherently contained and *expressed* the mechanisms of logic and Mathematics. "Do we give children the instruction 'think!' without even telling them how to think?" (p4, Papert's emphasis). The mathematical purpose expressed by Feurzeig is actually at odds with Papert who is at pains to stress the general problem solving capability of the language [17, 20]. Lisp's origins in Artificial Intelligence were supposed to support this [21 p14, 22 p16], but no author I have read ever explained how it was to happen.

5 Basic and Logo

Feurzeig's Logo group began with *education* and worked *back* to the form of their language. To create Basic Kemeny and Kurtz began with Computer Science and found educational uses for it. As someone who has spent forty years shouting at his education students to always begin with education and bring in computers if they could be useful I find it painful that Basic was an almost instant educational success and Logo wasn't.

By 1967 Basic was in use in eighteen secondary schools, eighteen colleges and universities, government agencies and "some local business concerns" [23 p23, 24 p2]. School use in particular was only limited by the number of available telephone lines. The reports are full of interesting and advanced programs written by students at all levels and the enthusiasm of the authors is obvious. (That said, Putnam, Sleeman, Baxter and Kuspa [25 p22] state "Errors were found with virtually every construct in all tests and interviews. ... students with a semester or more of experience with Basic had a very fuzzy knowledge of how various constructs operate.")

By contrast, many of the programs in the *Final Report on the Logo Project* [1], seem forced, elementary and repetitive. Many from the primary level, ages 7 to 9, are examples of programs to reverse the letters in a word, print a set of consecutive numbers or simply print strings. The first lessons did not involve writing code at all. This did not happen until lesson Seven (p. 67). In the secondary curriculum, many essential elementary functions such as divide and multiply were written by the teachers and given to the students to try and understand (p. 215), the inference being that they could not be expected to write these routines themselves. Johnson [26 p201] found "The position that the programming environments themselves, e.g., Logo microworlds, would become the school mathematics curriculum has clearly failed to gain the support of the educational system." (See also Mayer [27].)

None of this suggests a language easily taken up by beginners and used for their own purposes. Part of the reason has to be the use, initially, of recursion for all loops, definite or indefinite. Recursion is, as Papert has said repeatedly, a powerful problem solving tool [15, 17, 20, 28] and indeed it is. But then, so is calculus. Papert in particular has always insisted that Logo is designed to encourage experimentation, with students writing and testing their own creations. Papert worked with Jean Piaget for many years and passionately believes in the idea of 'learning by doing,' something he later extended to what might be termed 'learning by *making things*.'

Given this emphasis, it is difficult to understand the reliance on recursion at the expense of a general iterative statement. Not once in all my reading have I come across an assertion that students can be expected to discover a recursive solution to a problem on their own. All I can find are examples provided *to* students to explain, understand, and adapt. In his seminal book, *Mindstorms*, [15 p71] Papert states that "recursion stands out as the one idea that is particularly able to evoke an excited response." That might be so, but he devotes less than two pages to it, mentioning it once more in the Appendix in the context of 'circular logic' (p109). Brown and Rubinstein suggest that with suitable prior experience, students can write their own recursive routine to traverse a tree, but they give no clue to their success rate. They did find that "if a student couldn't figure out how to write a function, we could not slowly lead him down the path to discovery" [19: 43]. Once acquainted with WHILE—DO in Basic or Pascal, or even the primitive Dartmouth-Basic GOTO, students have no trouble in writing their own indefinite loops. (Murnane [29, 30]. See also Vitale [31 p272 & 272], and Murnane and Warner [32] for examples of experiments where children could have, but failed to use, recursion.)

A further reason which can be advanced is Logo's Lisp inheritance, essentially lambda calculus, whereas Basic was deliberately designed to be "as close to ordinary English combined with elementary algebra as possible" [23 p4]. Logo statements do not always accord with English, although in its earlier versions it approached it more closely than in later ones. Indeed, anyone coming to Logo after 1970 would be hard-put to recognise the original language. For example, Multiply [1 p215] is defined as:

```
TO MULTIPLY /X/ AND /Y/
10 IS /Y/ "0"
20 IF YES RETURN "0"
30 RETURN SUM OF /X/ AND MULTIPLY OF /X/ AND (DIFF OF
/Y/ AND 1)
END
```

Even allowing for the difficulty of conceptualising the recursion, it is not English, and in its early iterations, Logo struggled to make progress. The cure for many of these problems was provided by Seymour Papert.

Logo is often associated specifically with Papert, and particularly with his Turtle Graphics. He joined the project in January 1969 as a consultant [1 p1, 33] and his invention of the Turtle and its commands transformed the language.

A Turtle is a small robot which, when connected to a computer, can move and turn on the floor. At a stroke this eliminated the gap between entering a program and observing its outcome, since the Turtle could execute a command as soon as it was entered. It also solved the problem of the student understanding what the command did: they could walk around behind the Turtle following its movements with their own. While they might need to be taught the meaning of "TEST IS COUNT /SENTENCE/ 1 [18 p45] they could easily appreciate what FORWARD 100 meant because it accorded to their own body actions and their natural language. Turtle Geometry provides an immediate and meaningful environment for the beginner, relating body movement to the effect of a Logo statement. Papert [28 p24] describes this as the "idea of 'body-syntonic' representations of knowledge." (For a discussion of designing computer languages to correspond to natural language see Murnane [34]).

Along with Turtle commands came definite iteration: REPEAT :N, relieving the programmer of the need to write all loops recursively. A recursive loop can only be executed by writing a procedure and then executing it. In keeping with the idea of observing actions as the commands were entered, you could now type REPEAT 4 [FORWARD 100 RIGHT 90] and *watch* a square being drawn. Note also the close correspondence to English syntax.

Once the Turtle migrated from the floor to the screen, Logo became accessible and viable in any classroom.

Papert's other outstanding attribute was his ability as a teacher, educational theorist and writer. Probably no one has matched his output, or perhaps, his influence, on educational programming. His argument is that students "do not understand the kind of thing a mathematical structure is: they do not see the point of the whole enterprise" [15 p23]. By using Logo and the Turtle, these concepts can be made visible and concrete. Nevertheless, his pronouncements on the promise of Logo to help teach mathematical concepts [35 p3] such as angle, length, variables and differential geometry, as well as "epistemological primitives, such as the notion of a mathematical system itself" (p. 23) have not been supported by hard research. Ross and Howe [36 p147] found that "the research of the last decade into 'mathematics through programming' have been more encouraging than discouraging, but only mildly so."

Rather sadly, the weight of subsequent research suggests that programming in Logo, by itself, does not teach Mathematics. Students, unless specifically taught about these points, keep Logo and Mathematics entirely separate in their minds, and few teachers, perhaps persuaded by Papert, seem to do this, or do so with much success. An experiment with Year 8 students, all of whom had used Logo (in the form of LogoWriter), showed almost no traces of Logo when asked to perform tasks in which it could be expected to appear if Papert's theories are correct [32]. There were almost no signs of Logo functioning as a meta-language. Even Abelson, Barnberger, Goldstein and Papert [22 p10] rather sadly remark that "Logo did not succeed in displacing Basic as the almost universal computer language for schools."

6   Beyond the 60s

> At a minimum ... the teacher must be absolutely fluent in at thinking in Logo. Brown and Rubinstein [19: 4]
>
> Anybody who is at all serious about writing programs must avoid the temptation of thinking in a programming language. Juliff [37: 38]

These two quotations really summarise the different educational and practical orientations of Logo and Basic. Logo was intended to be used in the closed environment of education as a *language to think with*, and Basic was intended to introduce students to the world of programming. Kurtz [9 p3] insisted that "our one mistake was to include the word 'Beginners' in the name" but the *significant* letter in the title stood for *'All-purpose.'* Basic has been used for professional applications almost from its appearance. Logo has not, and was never intended to be.

At their inception, Basic and Logo were much simpler. Partly this was a result of choice by the designers: Basic was to be as easy as possible to learn and Logo was to encourage students to write most of their tools themselves, but there was also an element of 'the possible.' Kemeny and Kurtz were undoubtedly constrained by the small and slow machines they were working with and the need to build the system and compiler from scratch, in assembly language, with only themselves and their students as analyst/programmers. (They do characterise the GE-235 as "reasonably fast … with a 6 micro-second cycle time" [23 p5].) Logo, being developed slightly later, within a larger institution and on a single machine, probably suffered less in this respect, and certainly had the advantage of the system being written in Lisp, an existing, highly-adaptable, high-level language. I think it would be possible to make a strong argument that the simplicity of both should have been preserved, but the temptation to take advantage of developing Computer Science theory and faster machines probably made extension inevitable.

Both Basic and Logo have been extended far beyond the limits any of their creators could have imagined: the computer science of the 60's gave no inkling of the possibilities the personal computer and object-oriented programming would bring. Microsoft Visual Basic for instance, is to be found more in industrial applications that it is in a school. It is a graceful expansion, the language itself growing to include most of the trappings considered essential for complex and safe applications: declaration and strong typing of variables, procedures, explicit indefinite loops, implicit blocking and the incorporation of objects, but without losing its original form and flavour. Its syntax is still relatively straight-forward and is as suitable for education as the original.

Logo, in the form of MicroWorlds, is part of a full-blown multi-media/robotics environment and in 2010 is probably the only language that makes Cobol look small or offers the same invitation to write the same thing in so many different ways. Feurezig's successors seem to invent a new command every time they have a new idea, even when existing commands would seem to be perfectly suitable to the purpose. For instance, the Robotics version adds a completely new, quite separate set of commands to talk to the Logo RCX 'brick.' This leaves the existing, and quite adequate, 'Talkto' protocol in the main body of the language and separates Lego Robotics from the use of its rich array of logic. Redundancy in the language is therefore rife, while it is axiomatic in computer language design that there should only be one way to do something [38 p527]. (See Lindsey and van der Meulen [39 p174] for a particularly salutary example of the consequences of failing to observe this principle.) On the other hand, the MicroWorlds Backpack is a brilliant model of an object, thought the language itself cannot really be said to be object-oriented.

7 Conclusion

> We have found that the transition from Logo to Basic is fairly easy for most students, whereas the transition from Basic to Logo is often incredibly difficult. Brown and Rubinstein [19: 42]

Basic and Logo have both born out their creator's intention. Basic is easy to learn and has indeed become All-purpose. Logo has enriched countless classrooms and introduced students to new ways of thinking about, and expressing problem solutions.

Therefore Brown and Rubinstein, above, present a neat summation of the teacher's dilemma. Basic *has* proven to be easy to learn and get along with, but is not as likely to foster new ways of thinking and expression as Logo. This is Weyer and Cannara's point about "learning programming in full generality" [8 p3]. Does introducing students to Basic actually prevent, or mitigate against, as Brown and Rubinstein suggest, the development of a *full set* of programming tools? Logo's exponents insist it can do this, but it seems to be at the expense of leaving many students cold. Logo's advocates have not demonstrated the gains exposure to these ideas are supposed to bring. That said, given the enormous number of contributing factors, research demonstrating this is inherently extremely difficult, and the advocates of Basic have not done much better.

Essentially the teacher of today is in no better position than the pioneers, and is essentially dependent on *their own belief in the promise* that having students write programs will bring educational and other advantages. And because I am passionately in this school myself, I would be perfectly happy to do so in Logo or in Basic because I see the same promise in both, (though since I value simplicity, I would have a hankering after LogoWriter rather than the over-elaborated MicroWorlds). The pioneers of educational computing *knew* that programming a computer had educational benefits and set out to prove it, but I don't think the World listened.

References

1. Feurzeig, W., Papert, S., Bollm, M., Grant, R., Solomon, C.: Programming-Languages as a Conceptual Framework for Teaching Mathematics. Final report of the first fifteen months of the Logo project. Washington, D.C., Bolt, Beranek and Newman. R–1889, 329 p. (1969), ERIC ED 007 932
2. Ham, V.: Technology as Trojan horse. In: McDougall, A., Murnane, J.S., Jones, A., Reynolds, N. (eds.) Researching IT in Education Theory, Practice and Directions, pp. 25–38. Routledge, London (2010)
3. Kemeny, J., Kurtz, T.: The Dartmouth Time-Sharing System, 76 p. National Science Foundation, Washington (1967)
4. Cubin, L.: Oversold and underused: computers in the classroom. Harvard University Press, Cambridge (2001)
5. Cox, M.: The changing nature of researching IT in education. In: McDougall, A., Murnane, J.S., Jones, A., Reynolds, N. (eds.) Researching IT in Education Theory, Practice and Directions, pp. 11–24. Routledge, London (2010)
6. Munro, R.K.: Setting a new course for research on information technology in education. In: McDougall, A., Murnane, J.S., Jones, A., Reynolds, N. (eds.) Researching IT in Education Theory, Practice and Directions, pp. 46–53. Routledge, London (2010)
7. Ershov, A.P.: Programming: the second literacy. In: Lewis, R., Tagg, D. (eds.) Computers in Education, vol. 1, pp. 1–8. North-Holland, Amsterdam (1981)
8. Weyer, S.A., Cannara, A.B.: Children learning computer programming: experiences with languages, curricula and programming devices. Stanford, Calf., Stanford University, 228 p. Technical Report No. 250 (1975), ERIC ED 111 347
9. Kurtz, T.: Basic. In: Wexelblat, R.L. (ed.) History of Programming Languages. Academic Press, New York (1981)

10. Papert, S.: Talking Turtle (Videorecording). Open University Educational Enterprises, Holt-Saunders, England (1993)
11. Dwyer, T.A.: Teacher-Student Authored CAI Using the NEWBASIC/CATALYST System, p. 22. Pittsburgh Univ., Pa, National Science Foundation, Washington, DC (1970), ERIC ED 043 235
12. Dwyer, T.A.: Some principles for the human use of computers in education. International Journal of Man-Machine Studies 3, 219–239 (1971)
13. Dwyer, T.A.: Teacher/Student authored CAI using the NEWBASIC system. Communications of the ACM 15, 21–27 (1972)
14. Dwyer, T.A.: Heuristic strategies for using computers to enrich education. International Journal of Man-Machine Studies 6, 137–154 (1974)
15. Papert, S.: Mindstorms: children, computers, and powerful ideas. Basic Books, New York (1980)
16. di Sessa, A.A.: Reflections on component computing from the Boxer project's perspective. Interactive Learning Environments 12, 161–165 (2004)
17. Papert, S.: Teaching Children Thinking, 241 p. Massachusetts Institute of Technology, Cambridge (1971), ERIC ED 077 241
18. Feurzeig, W., Kukas, G., Faflick, P., Grant, R., Lukas, J.D., Morgan, C.R., Weiner, W.B., Wexelblat, P.M.: An Introductory LOGO Teaching Sequence: LOGO Teaching Sequence on Logic, 329 p. Beranek and Newman, Cambridge (1971), ERIC ED 057 579
19. Brown, J.S., Rubinstein, R.: Recursive functional programming for the student in the humanities and social sciences, p. 53. University of California, Irvine, UCI–ICS–TR–27a (1974), ERIC ED 108 664
20. Papert, S., Solomon, C.: Twenty things to do with a computer, 240 p. Massachusetts Institute of Technology, Massachusetts (1971), ERIC ED 077 240
21. Evens, P.: What is Logo? Deakin University Press, Geelong (1992)
22. Abelson, H., Barnberger, J., Goldstein, I., Papert, S.: Logo progress report 1973-1975, p. 22. Massachusetts Institute of Technology, Cambridge, AI Memo 356 (1976), ERIC ED 128 181
23. Kemeny, J., Kurtz, T.: The Dartmouth Time-Sharing System, 76 p. National Science Foundation, Washington (1967)
24. Kurtz, T.: Demonstration and Experimentation in Computer Training and Use in Secondary Schools, Activities and Accomplishments of the first year, 81 p. Hanover, Dartmouth College (1968), ERIC ED 027 225
25. Putnam, R., Sleeman, D., Baxter, J.A., Kuspa, L.K.: A summary of misconceptions of high school Basic programmers. Stanford University School of Education, Stanford, Occasional Report #010 (1984), ERIC ED 258 556
26. Johnson, D.C.: Algorithmics and programming in the school mathematics curriculum: support is waning—is there still a case to be made? Education and Information Technologies 5, 201–214 (2000)
27. Mayer, R.E.: Introduction to research on teaching and learning computer programming. In: Mayer, R.E. (ed.) Teaching and Learning Computer Programming, pp. 1–12. Lawrence Erlbaum, Hillsdale (1988)
28. Papert, S.: The Turtle's long slow trip: Micro-educational perspectives on Microworlds. Journal of Educational Computing Research 27, 7–27 (2002)
29. Murnane, J.S.: Models of recursion. Computers & Education 16, 197–201 (1991)
30. Murnane, J.S.: To iterate or to recurse? Computers & Education 19, 387–394 (1992)
31. Vitale, B.: Elective Recursion: A Trip in Recursive Land. New Ideas in Psychology 7, 253–276 (1989)

32. Murnane, J.S., Warner, J.W.: An empirical study of secondary students expression of algorithms in natural language. In: McDougall, A., Murnane, J.S., Chambers, D. (eds.) 7th IFIP World Conference on Computers in Education. Computers in Education 2001: Australian Topics, vol. 8, pp. 81–86. Australian Computer Society, Bedford Park (2001)
33. Davis, R.B.: Editorial. The Journal of Mathematical Behavior 10, 4 (1991)
34. Murnane, J.S.: The psychology of computer languages for introductory programming courses. New Ideas in Psychology 11, 213–228 (1993)
35. Papert, S.: Teaching children to be mathematicians vs teaching mathematics, p. 26. Massachusetts Institute of Technology Artificial Intelligence Laboratory, Cambridge (1971), ERIC ED 077 243
36. Ross, P., Howe, J.: Teaching mathematics through programming: ten years on. In: Lewis, R., Tagg, D. (eds.) Computers in Education, vol. 1, pp. 143–148. North-Holland, Amsterdam (1981)
37. Juliff, P.: Programming—should we enjoy it or do it properly? In: Welch, R. (ed.) Ninth Australian Computer Conference, pp. 38–43. Mercury-Walch, Hobart (1982)
38. MacLennan, B.J.: Principles of Programming Languages. Holt, Rinehart & Winston, New York (1983)
39. Lindsey, C.H., van der Meulen, S.G.: An Informal Introduction to Algol 68. North Holland, Amsterdam (1973)

The Life and Growth of Year 12 Computing in Victoria: An Ecological Model

Arthur Tatnall[1] and Bill Davey[2]

[1] Graduate School of Business, Victoria University, Australia
Arthur.Tatnall@vu.edu.au
[2] School of Business Information Technology, RMIT University, Australia
Bill.Davey@rmit.edu.au

Abstract. This paper seeks to apply the techniques of ecology, used in a wide range of fields, to analyse a series of events in the history of computing. The case analysed here is the history of development of computer studies curriculum at the senior level in high schools in the Australian state of Victoria. Although theoretically directed by a central body, development of the curriculum for the final high school year in computing shows a history containing many anomalies. Applying an ecological method to the historical narrative shows that seemingly illogical changes can be explained by the interaction of organisms and the environment in which the history has taken place. From this example we will show that ecological principles can also be useful in analysing an historical event.

Keywords: Computer Science, Information Technology, computer studies curriculum, history, senior secondary school curriculum, ecological methods.

1 Introduction

In many school systems the 'worth' of a subject is measured in terms of its interaction with the final year of schooling (Year 12) as many ascribe special importance to those subjects that are seen as valid preparations for tertiary study. In 1981, as a result of many years of effort by a group of academics, *Computer Science* was first offered as a Higher School Certificate (Year 12) subject in the Australian state of Victoria [1].

Until the 1960s the final year of high school in Victoria had been known as Matriculation. This word, meaning qualification to enter the University, illustrates the way in which the final year of high school was seen as a preparation for university studies, and all subjects were closely aligned to a similar study at university. In the 1960s in a review of the senior secondary school system the Higher School Certificate (HSC) was introduced in place of the Matriculation certificate. Another review of upper secondary school curriculum in the early 1990s replaced the Higher School Certificate by the Victorian Certificate of Education (VCE) that extended over Years 11 and 12.

Introduction of the VCE brought with it a number of changes to existing subjects, and saw *Computer Science* replaced with three new Year 12 subjects: *Information Processing and Management*, *Information Systems* and *Information Technology in Society*. Each of these subjects comprised two units. An additional new subject: *Information Technology* was offered only at Year 11.

A. Tatnall (Ed.): HC 2010, IFIP AICT 325, pp. 124–133, 2010.

2   An Historical Approach

Some writings in the history of computing have sought to record, in an uncensored narrative, the stories of the pioneers of computing (before we lose them). On the other hand there is also an opportunity for articles to analyse the history presented in these narratives and seek to learn from the trends that are uncovered. George Santayana is often quoted as saying "If we do not learn from the mistakes of history, we are doomed to repeat them." His philosophy text [2] actually said "Those who cannot remember the past are condemned to repeat it", but in either case it is the analysis and historical record that allow us to avoid condemnation, or the repetition of mistakes.

Historical record is a complex field and in the analysis of complex fields a number of techniques have been developed to avoid the perils of reductionism. Often in applying scientific method to complex situations we can remove meaning at the same time that we remove complexity. In this paper we will take a particular piece of the history of computing and apply an ecological method to draw out the meaning contained in the history.

The science of Ecology has developed many different ways of investigating nature. Ecology is concerned with interrelationships between different living things, and between living things and their environment [3]. In addition to dealing with the natural environment however, the principles of Ecology have been used in many other areas to deal with the complexity of those areas [4-13]. These techniques have also been used in education and curriculum development to produce worthwhile results [3, 14-19]

2.1   Anomalies in the Historical Case

In this paper we use as the unit of analysis, the history of the development of computing studies curricula in high schools in Victoria. In Victoria the senior secondary school curriculum is controlled by a central body that has had a number of names during the history we are recording, but is currently called the Victorian Curriculum and Assessment Authority. This authority attempts to respond to the needs of the community to produce new HSC subjects as the need arises. These subjects are reviewed on a regular basis and changes are made in response to industry, community and university suggestions. University entrance in Australia is highly competitive and determined almost completely by the results obtained by the student in their final year of high school. There are state-wide and nationally moderated assessments of all final high school subjects to determine a score (called the ENTER score) used by most universities for entrance requirements. This means that the authority must make changes to the curriculum carefully in logical response to real needs.

Despite this central control, the intense pressure to make sure that all subjects offered meet the needs of society and students, and attempts to make only rational decisions in terms of this curriculum, the history of the development of computing curriculum is filled with anomalous outcomes. These include:

- One subject introduced in order to make computing accessible and popular has been abandoned due to low enrolments.

- Changes to other subjects intended to make them more popular with girls have seen little in the way of improving the long-term gender balance.

3 Elements of the Ecological Method

Townsend, Harper and Begon [20] suggest that two key biological principles exemplify the concepts of ecology:

- Organisms behave in ways that optimise the balance between their energy expenditure and the satisfaction they obtain.
- Organisms operate within a competitive environment that ensures only the most efficient of them will survive [3].

Using the principles of ecology to analyse complex situations means more than just trying to find a correspondence or a metaphor between the situation and the ecological entity. In ecology it has been found useful to make distinctions between organisms, resources and the environment in which these organisms must prosper or die, and there are simple principles that underlie the success or failure of organisms within an environment. Habitat, ecological niches, the exploitation of resources in predator-prey interactions, competition, and multi-species communities [21, 22] are all important considerations in ecology [3]. These principles, however, are also found to be useful in analysing other complex situations that do not necessarily involve biological entities. The first step in an ecological analysis requires us to identify several different entities:

- Those things that cannot be changed by the organisms (actors) we will call the **environment**.
- Those things that are consumed by the organisms but are unable to influence these organisms, except by their consumption, we will call **resources**.
- Those things that are able to consume resources and interact with each other, and which may prosper or die over time we will call the **organisms**.

When seeking to explain the success or failure of an organism we will look for interactions between the organism, its environment and the resources available. There are several principles found in biologically ecologies that are common to other complex situations:

- Organisms can prosper if they find a **niche** in the environment that particularly suits them.
- Organisms prosper when they use the **least energy** to obtain the greatest response.
- Organisms can prosper by being **cooperative** with other organisms in the environment.
- Organisms can prosper through **competitive behaviour** with other organisms and environments.

An ecosystem is a highly complex entity due to the large number of living things inhabiting it, and to the variety of interactions possible between each of these [15]. We will look for these ecological principles within the history of our case.

4 Analysing This Case Using the Ecological Model

In this case the environment includes: University entrance places – these are a rare and difficult location to reach in the environment, Government policy and public perceptions. Environmental resources include: student enrolments – these are attracted by the different subjects that make up the final year of high school. A subject that is able to attract a large number of students will be successful, while one unable to attract students will die.

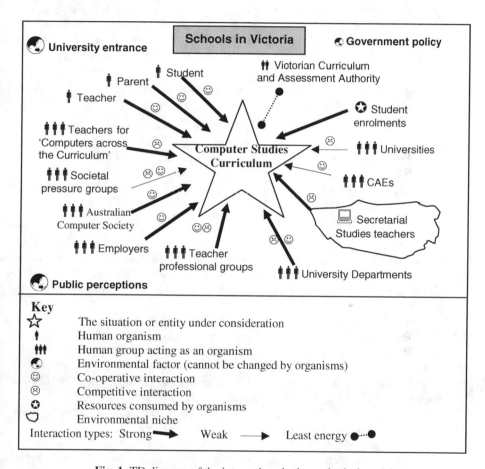

Fig. 1. TD diagram of the interactions in the ecological model

Organisms include: the subjects – these are non-biological organisms, students, teachers, universities, industry, employers, the Australian Computer Society (ACS), pressure groups and the general public. University departments seek to support subjects that they see as co-operating with them. Teachers represented through their professional groups seek to make changes in subjects – subject-based teacher groups support their own subjects.

We will now look at the way the history of this development unfolds from the point of view of the ecological model.

4.1 Turning Points and Ecological Explanations

4.1.1 The First Turning Point We Examine Is the Birth of a New Organism

In the Victorian education system, as in many education systems around the world, senior high school studies have closely reflected the major disciplines in universities. When the HSC subject *Computer Science* was first introduced in 1981, however, it was done without the existence of many corresponding subjects in Australian universities.

Our ecological model asks us to look at the relationships of this new subject to the other organisms and to the environment: tertiary institutions, schools, teacher groups and the general public. Most of the pressure to introduce this subject came from outside the Education Department, and secondary school teachers had little involvement in determining its nature or content.

> *The public: parents, students and employers; readily accepted HSC Computer Science, and student numbers taking the new subject increased rapidly. In its first year, 120 students from 10 schools studied Computer Science, but this number increased by over 50% in each of the next five years before making a slower increase to over 2200 in 1991 (only a little smaller than Geography) with virtually every secondary school in the state offering the subject.* [23]

Melbourne and Monash universities, which saw themselves as guardians of academic standards, initially rejected the subject, not allowing its inclusion in admission scores for their courses. Their stated reason for this was that the component of assessment allotted to the formal examination was only 35% in HSC Computer Science rather than the usual 50%. (The subject designers had allotted only 35% to the formal examination to allow space for practical work in programming to be given some weight.) When pressed, several academics from these institutions admitted that they considered the subject of little serious academic worth, and 'not an appropriate subject to study at a secondary school level' [1]. They would often state that they would prefer to have a student with no knowledge of computing who they could introduce to computing concepts, rather than one 'who had learned bad programming in BASIC'. The fact that the Computer Science subject had been specifically constructed to address this particular concern by the study of top-down design and structured programming missed their notice.

Parallels can be drawn between the introduction of Computer Science and that of Geography in English schools as described by Goodson [1, 24]. In his analysis, Goodson makes use of Layton's [25] three stage model of the growth of science education in nineteenth century England.

1. Layton suggested that in the First Stage, a new subject stakes its place based on grounds of pertinence, utility and the needs and interests of the learners. In this stage learners are attracted as the subject relates to matters that concern them. Teachers tend not to be trained specialists, but pioneers who bring missionary zeal to their task.

2. In the Second Stage a tradition of scholarly work begins to emerge and more teachers are recruited as trained specialists. Students are now increasingly attracted by the new subject's reputation and growing academic status. Subject matter in the new subject is becoming increasingly well organised.
3. In the Third and final stage the teachers have constituted a professional body with established rules and values. Subject matter is largely determined by specialist scholars in the field and students are thus initiated into a tradition.

An interesting variation is the reaction to the subject by tertiary institutions in Victoria. Several of the 'traditional' universities, which had been offering a highly abstract version of Computer Science for some time opposed introduction of the new subject at secondary school level, claiming that the concepts required of this discipline were beyond what could be understood by 17 year old students. On the other hand many of the newer tertiary institutions, then known as Colleges of Advanced Education (later to become universities), were generally quite supportive after fighting their own battles with the universities on the place of computing some years earlier. This partnership can be seen as a co-operation between the organism represented by the new subject and the organism represented by these new tertiary institutions.

Parents, students and employers readily accepted HSC Computer Science, and student numbers taking the new subject increased rapidly. Teachers, however, were not universally in favour. Beginning from about the mid-1980s when Computer Science was still in rapid growth a number of teachers began to question its place [1]. Their argument had several strands. Firstly, some claimed that HSC Computer Science was an elitist academic subject, too difficult for some students, and so should not be supported (- it is interesting to note that this was exactly opposite to the view earlier expressed by Melbourne and Monash universities). Others noted that the ratio of girls to boys taking Computer Science was almost as low as that for physics, and expressed concern that it was becoming a *boy's subject*. Perhaps, however, the most damaging criticism came from those teachers who claimed that the presence of a specialist subject detracted from the move to encourage the use of '*Computers across the Curriculum*'. Their argument had two parts: firstly, they argued that the demands made on school computing facilities by Computer Science classes made it difficult for others to obtain adequate access to the machines. While containing some truth, this argument misses the point that in most cases the reason that the school had purchased a number of computers at all was to support the teaching of Computer Science. Secondly, it was argued that the existence of a specialist subject would mean that teachers of other subject areas would not bother to include any mention of computing, considering it covered elsewhere. A number of teachers saw Computer Science and Computers across the Curriculum as adversaries and it took some time before these competing points of view were reconciled [1].

4.1.2 The Second Turning Point: Three New Subjects
Changes to the structure of the Year 12 curriculum in the early 1990s resulted in the introduction of three new Information Technology subjects: *Information Systems*, *Information Processing and Management* and *Information Technology in Society* to replace *Computer Science*. This process also resulted in a decrease in any links between secondary curriculum and the universities. At this time several other subjects

were made redundant as they were seen as having 'died' due to a perceived lack of interest. A subject of particular significance here was *Secretarial Studies* which had occupied a comfortable niche as a vocational subject preparing (mostly) girls for entry into office and administrative jobs. The idea of teaching typing on typewriters was, however, seen as anachronistic. The numerically large and influential group of teachers made redundant in this process produced a pressure group. These teachers had been moving slightly in their niche towards the use of electronic typewriters. This pool of teachers threatened with significant changes to their work on the demise of their own teaching subject can be seen as a powerful predator. When Secretarial Studies teachers were forced to move into computing, some attempted just to move their previous niche into a new location by ignoring, or de-emphasising, any content material that did not suit them. This aspect of their move into teaching Information Technology subjects can still be seen in the nature and scope of computing studies in the final school year.

Another predator within the environment was a group of teachers with a philosophical view that computers should be a tool used in every part of the curriculum *rather* than a topic to be studied by in its own right. After their success in killing off Computer Science, this 'Computers across the Curriculum' group continued to influence the way computers were used in secondary schools; an interesting example being that of mathematics. In mathematics every year level was now required to add computer-based learning of some kind into the curriculum. In 1999 it was found that computing resources in schools were not up to this demand, and a trial of using computers in the examination system failed. Mathematics teachers found the solution to this problem in using relatively inexpensive graphics calculators rather than computers, and use of Excel and Mathematica in schools decreased to almost zero.

4.1.3   The Third Turning Point: Death of One of the New Subjects

Another anomalous thing that happened was the downward progress of the subject: *Information Technology in Society*. By 1997 the numbers in this subject had become so low that it was not re-accredited for the next year. This is seen as an anomaly as the creation of the new courses in information technology was based on a few simple, logical propositions.

Information Technology was seen as having three main areas of interest:

1. Studying information technology itself – including programming
2. Studying the uses of information technology – especially in business
3. Studying the wider impact of information technology across the whole social landscape.

The idea of three streams was to cater to a wide audience. It was presumed that students fell into interest groups including those interested in technology and those interested in the social sciences. To have one of these streams collapse in such a short period of time showed that this model of students logically following some stereotyped interest is not valid.

In ecological terms we find a number of principles in action in this turning point. The creation of the stream is an example of the least energy relationship between the curriculum authority and subject offerings. A subject was proposed by the subcommittee handling computer related subjects and the option with the greatest return on energy expenditure for the authority was to offer the subject and see if it captured sufficient student numbers to survive. The subject had little chance to succeed in this environment as there was a co-operative relationship between university departments and both the other streams, but no interest from any department in the IT and Society subject. Similarly the industry groups anxious to increase the number of available employees in computing had no interest in co-operating with a subject that covered the potential evils of technology without developing many job related skills.

4.1.4 The Fourth Turning Point: Failure of the New Subjects to Attract Girls

One of the principal tenets for creation of the change from Computer Science to the three new streams of computing (Information Systems, Information Processing and Management and Information Technology in Society) was the desire to make the subjects more interesting and attractive to girls. By 2008, however, we find the gender balance in the two remaining subjects (Information Systems, Information Processing and Management) to have dropped to less than 25% girls. Clearly the intention to make these subjects amenable to girls has not been achieved and we need to look into the history of computing subjects to try to find out why.

The ecological approach is required to understand this continued failure, despite deliberate and conscious efforts to change subjects to include material of interest to girls. In this case the important issue is competitors in the form of other HSC subjects. The enrolment of females in the final year of high school and in subsequent university places increased over this period. These student enrolment resources were typically being taken up by 'traditional' female subjects such as those in commercial streams including legal studies, accounting and such, but were also being increasingly captured by biology and even mathematics. The commerce subjects can be seen as an example of least energy expenditure: a female student choosing a subject because she can 'end up doing the same things that mother did'. The movement towards biology and mathematics appears to be related to the large efforts in marketing by the feminist lobbies in society. These teacher and university department based groups were able to see a co-operative relationship between their aims and the identified subjects. This co-operative relationship was, however, never developed with any of the information technology subjects.

5   Conclusion

A naïve view of curriculum development sees central organisations making changes to curriculum in high schools in logical response to the needs of industry and society. The most cursory examination of any example of the history of computing in schools, however, shows numerous examples of changes that are neither logical nor related to changes in industry or society. Often central bodies are required to cancel subjects because of the lack of student numbers. We have shown here, using a particular case, that the history of computing in schools can be understood more clearly by using an ecological model based on the organisms, resources and the environment and looking at their interactions in terms of niches, expenditure of the least energy, cooperative and competitive behaviour.

References

1. Tatnall, A.: The Growth of Educational Computing in Australia. In: Goodson, I.F., Mangan, J.M. (eds.) History, Context, and Qualitative Methods in the Study of Education, pp. 207–248. University of Western Ontario, Canada (1992)
2. Santayana, G.: The Life of Reason, vol. 1. C. Scribner's Sons (1905)
3. Tatnall, A., Davey, B.: Improving the Chances of Getting your IT Curriculum Innovation Successfully Adopted by the Application of an Ecological Approach to Innovation. Informing Science 7(1), 87–103 (2004)
4. Star, S.L., Griessemer, J.R.: Institutional Ecology, 'Translations' and Boundary Objects: Amateurs and Professionals in Berkley's Museum of Vertebrate Zoology, 1907-39. Social Studies of Science 19, 387–420 (1989)
5. Grzywacz, J.G., Fuqua, J.: The Social Ecology of Health: Leverage Points and Linkages. Behavioral Medicine 26(3), 101–115 (2000)
6. Barnett, W.P., Mischke, G.A., Ocasio, W.: The Evolution of Collective Strategies among Organizations. Organization Studies 21(2), 325–354 (2000)
7. Simon, D.: Ecological Metaphors of Security: World Politics in the Biosphere. Alternatives: Social Transformation & Humane Governance 23(3), 291–320 (1998)
8. Sutcliffe, A., Chang, W.-C., Neville, R.: Evolutionary Requirements Analysis. In: 11th IEEE International Requirements Engineering Conference (RE 2003). IEEE, Monterey Bay (2003)
9. Podolny, J.M., Stuart, T.E.: A Role-Based Ecology of Technological Change. American Journal of Sociology 100, 1224–1260 (1995)
10. Havelka, D., et al.: Evolution of IS Professionals' Competency: an Exploratory Study. Journal of Computer Information Systems 41(4), 21 (2001)
11. Johnston, R.: Panel: Evolution of Computing in Spanish Speaking Countries (IFIP WCC 2006). IFIP, Santiago (2006)
12. Nagarajan, A., Mitchell, W.: Evolutionary Diffusion: Internal and External Methods Used to Acquire Encompassing, Complementary, and Incremental Technological Changes in the Lithotripsy Industry. Strategic Management Journal 10(11) (1998)
13. Richards, R.M., Sanford, C.C.: An Evolutionary Change in the Information Systems Curriculum at the University of North Texas. Computers and Education 19(3), 219–228 (1992)
14. Tatnall, A., Davey, B.: Curriculum Development in the Informing Sciences: Ecological Metaphor, Negotiation or Actor-Network? In: Informing Science and IT Education Conference. University College Cork, Cork (2002)
15. Tatnall, A., Davey, B.: Information Systems Curriculum Development as an Ecological Process. In: Cohen, E. (ed.) IT Education: Challenges for the 21st Century, pp. 206–221. Idea Group Publishing, Hershey (2002)
16. Tatnall, A., Davey, B.: Understanding the Process of Information Systems and ICT Curriculum Development: Three Models. In: Brunnstein, K., Berleur, J. (eds.) Human Choice and Computers: Issues of Choice and Quality of Life in the Information Society, pp. 275–282. Kluwer Academic Publishers/IFIP, Assinippi Park (2002)
17. Tatnall, A., Davey, B.: ICT and Training: A Proposal for an Ecological Model of Innovation. Educational Technology & Society 6(1), 14–17 (2003)
18. Tatnall, A., Davey, B.: A New Spider on the Web: Modelling the Adoption of Web-Based Training. In: Nicholson, P., et al. (eds.) E-Training Practices for Professional Organizations, pp. 307–314. Kluwer Academic Publishers/IFIP, Assinippi Park (2005)

19. Tatnall, A., Davey, B.: Information Systems Curriculum Using an Ecological Model. In: Khosrow-Pour, M. (ed.) Encyclopedia of Information Science and Technology, 2nd edn., pp. 1998–2003. Idea Group Reference, Hershey (2009)
20. Townsend, C.R., Harper, J.L., Begon, M.: Essentials of Ecology. Blackwell Science, Boston (2000)
21. Case, T.J.: An Illustrated Guide to Theoretical Ecology. Oxford University Press, New York (2000)
22. Krebs, C.J.: Ecology - The Experimental Analysis of Distribution and Abundance, 3rd edn. Benjamin Cummings, San Francisco (2001)
23. Tatnall, A.: Curriculum Cycles in the History of Information Systems in Australia. Heidelberg Press, Melbourne (2006)
24. Goodson, I.F.: School Subjects and Curriculum Change: Studies in Curriculum History - A Revised and Extended Edition. The Falmer Press, UK (1987)
25. Layton, D.: Science as General Education. Trends in Education (January 1972)

History of the European Computer Driving Licence

Denise Leahy and Dudley Dolan

School of Computer Science and Statistics, Trinity College, Dublin, Ireland
{denise.leahy,dudley.dolan}@scss.tcd.ie

Abstract. The European Computer Driving Licence (ECDL) began as a project which set out to define the computer skills required by the ordinary citizen to take advantage of the new end user technology. The project started in 1995 and since that time ECDL has become the leading digital literacy certification in the world with almost 10 million candidates enrolled in the programme. The core ECDL consists of seven modules, defined by a syllabus which is agreed by an international panel of users and experts and certifies that the holder has the competencies required to perform basic tasks using a personal computer, can use a computer in practice and understands the basic concepts of information technology [1]. This paper describes the creation of the ECDL, together with the unique organizational structure which enabled the wide implementation of ECDL, initially in Europe and later throughout the world as the International Computer Driving License (ICDL).

Keywords: Digital Literacy, Certification, IFIP, CEPIS, Basic ICT Skills.

1 Introduction

1.1 The Initial Idea

With the advent of home computing and the growth in the general use of computers in the early 1990s, the Council of European Professional Informatics Societies (CEPIS) recognised the importance of defining the computer skills required by the ordinary citizen to take advantage of the new technology. The idea originated in Norway and preliminary investigations were undertaken to see what activities were being carried on in this area. The idea was supported by the European Commission and the member societies of CEPIS. CEPIS created a task force in 1995 to examine how to raise the level of such skills in industry throughout Europe [2]. Initially the CEPIS task force consisted of the Nordic countries plus Ireland (Norway, Finland, Sweden, Denmark, and Ireland.) These countries met in 1995 and, after the initial findings were reported to the CEPIS Council, a number of other countries decided to join an extended task force. The extended task force consisted of the original countries plus the new representatives from Austria, France, Italy, The Netherlands and United Kingdom.

This paper describes the history of ECDL from the setting up of the Task Force in 1995 through to the development of ECDL in Europe and later throughout the world. Using material and working papers from that time, it describes the development of

A. Tatnall (Ed.): HC 2010, IFIP AICT 325, pp. 134–145, 2010.

ECDL, the major milestones and the support given initially by the European Commission and later by Governments, industry and educational bodies.

1.2 The Links between CEPIS and IFIP

The International Federation for Information Processing (IFIP) is a non-governmental, non-profit umbrella organization for national societies working in the field of information processing. IFIP was established in 1960, under the auspices of UNESCO, as a result of the first World Computer Congress held in Paris in 1959. This provided a forum for persons involved in the Information Technology area to meet and share ideas [3]. In the 1980s, some of the IFIP member societies based in Europe felt that there was a need for a European organization to meet the local needs of the region and CEPIS was established in 1989 by 8 European informatics societies, CEPIS has since grown to represent over 300,000 informatics professionals in 33 countries. There are many common member societies in IFIP and CEPIS and they communicate and share research and interests at their respective events and meetings [4].

The first area of focus for CEPIS was the promotion and development of information and communications technology (ICT) skills across Europe. CEPIS set out to work closely with the European Commission in order to address the needs of the European Community and meetings were held with Directorates General in the European Commission to establish the areas which CEPIS should pursue. An early venture for CEPIS was the European Informatics Skills Structure which set out to define the skills for ICT professionals in a structured format which could be adopted throughout Europe [5]. This work commenced with the founding of CEPIS in 1989 and work continued until 1996. The project which created ECDL started in 1995.

2 The Information Society and Digital Literacy in the 1990s

In 1990, 23 million European households owned a PC and this rose to over 40 million in 1994. However the proportion of European households with a PC was significantly lower than the USA and there were fewer multi-media PCs or those linked to a communications network. Statistics also showed that the speed of adoption of the new technologies highlighted cultural and linguistic differences between Northern and Southern Europe and between the US and Europe [6]. At that time in the United States there were 34 PCs per hundred citizens with the European figure at 10 per hundred.

Advances in information technology substantially changed the workplace in Europe in the mid-nineties. Over 72% of office workers had a PC or equivalent on their desks and used it as part of their work. At that time there was concern about the future of employment, skills and technology. According to Cortada in 1998 "Each year, on average, more than 10% of all jobs disappear and are replaced by different jobs in new processes, in new enterprises, generally requiring new, higher or broader skills. There is a much slower pace on the supply side in the acquisition of new skills." [7] He suggested that within 10 years people in employment would be using new technology with out of date skills.

The area of unemployment was recognized as of vital importance, creating a need for retraining. The European Union 1996 report "Living and working in the

Information Society" advised that "instead of having 9 million people in long term unemployment and de-skilling, the most expensive form of public spending with the lowest return to the economy or the individual, and many more millions on their way to long term unemployment, the Member States of the EU should have 9 million involved in upgrading, maintaining and improving their skills in literacy, numeracy and IT." [8].

In the '90s there was much discussion about the new "Information Society" and the different skills needed by all individuals to take full benefit of what the new society could offer. It was recognized that the new Information Society in Europe would affect everyone. The EU Commissioner Martin Bangemann noted in "The Information Society and the Citizen" that we needed to make "greater efforts" in our schools to prepare the next generation to participate and benefit fully from the Information Society and to stimulate European citizens to create new services in education, entertainment and business in order to keep Europe at the forefront of technology [9].

It was clear that there was a need but it was not clear how to define this. At this time there was talk of computer literacy, digital literacy, end user skills and other definitions. In 1997, Paul Gilster suggested that to be digitally literate a person should be able to find information on line in web sites, databases and other on line information resources, be able to evaluate that information and use email and search engines [10].

It was part of the vision of the CEPIS task force to address the issues raised above and contribute to the successful development of an all inclusive Information Society.

3   The Beginning

The Finnish Computer Driving Licence (CDL) was introduced in Finland in January 1994 with the support of the Ministry of Education, the Central Organisation of Finnish Trade Unions, the Confederation of Finnish Industry and Employers, the Finnish Information Processing Association and the Ministry of Labour [11]. The first Computer Driving Licences were awarded in early 1994 and some 10,000 had been issued by December 1995. In 1995, the Finnish Information Processing Association (FIPA) brought the CDL to CEPIS as a potential model for a wider European context.

The CEPIS task force (named the User Skills task force) met in 1995 to consider how to increase the competence required for the European work force and for the European individual. This task force looked for a suitable model and examined the Finnish CDL in detail. There were other models in Europe at that time and these were also examined. After thorough study it was concluded that the basic Finnish concept was widely applicable throughout Europe. However, changes and updates were required – the Finnish model consisted of a bank of questions which they called the "question and test base" but had no defined syllabus [1].

In order to assess the modifications needed to have the ECDL meet the requirements of a wider marketplace, a series of pilot tests were carried out in Norway, Sweden, Denmark, France and Ireland. As a result of these tests and a thorough evaluation of the concept, modifications were agreed. It was decided to name the concept "the European Computer Driving Licence" or "ECDL".

The pilot tests gave an opportunity to evaluate the concept from a number of viewpoints. The pedagogical aspects were considered from the point of view of the breadth of knowledge required and the depth needed. The need for the theoretical module was considered and it was decided to retain this. A syllabus was created and a new question and test base was developed to meet these newly defined requirements.

In addition to evaluating the tests themselves, this period of time was used to establish what the market response would be to such a concept. The response was uniformly optimistic although the numerical forecasts varied considerably due to the different levels of penetration of PCs in the different countries. It was clear from the research that industry felt a need for some sort of certification to ensure that their investment in training was worthwhile. Individuals welcomed the opportunity to show that they had acquired computer skills which could be certified. Potential course vendors and training organizations were happy with the concept as it gave them a focus for their course developments. Governmental and employer organizations and Trade Unions in many countries found the concept attractive. In particular, the Irish Congress of Trade Unions (ICTU) was very supportive at a general meeting in Dublin in March 1996 [12]. Ministries of Education in a number of countries gave active support. In Denmark a Government supported regional initiative was undertaken.

The User Skills task force created a syllabus, tests, guidelines and procedures to run ECDL on a European scale. The ECDL certification developed by the task force was a definition of the competencies required to perform basic tasks using a personal computer in order to certify that the holder had the skills required to use a computer and had some theoretical knowledge of computers. The ECDL, similar to the Finnish model, had seven modules. The first ECDL syllabus was published in October 1996 and this represented one of the main outcomes of the work of the ECDL Task Force during 1996. It was expected that the syllabus would be updated on a continual basis and new versions would be issued every year.

Individuals were to prove their proficiency by passing a test within each module with progress registered on a Computer Skills Card. An ECDL was awarded once all seven module tests were passed successfully. Except for one theoretical module, the tests were skills oriented. The tests were independent of both machine and software vendors and products. The infrastructure of CEPIS member societies as licensees to run ECDL within a country was set up. The procedures and rules for setting up Test Centres and for administering, running and marking tests in a standard and consistent way were devised. The legal and contractual issues were identified and put in place.

An initial method for administering the tests automatically by computer was researched in Sweden and this was extensively evaluated. It satisfied many requirements but was not seen to be the final solution to the automation needs in Europe at that stage. This was developed further at a later stage.

4 ECDL 1997 - 2010

The European Computer Driving Licence had a number of clearly identified objectives in 1995 as follows [13]:

- To raise the level of IT competence within the work force in industry, commerce and public services throughout Europe.
- To provide a basis for certifying computer skills in all levels of the education sector and provide a basis for certification of skills for lifelong learning.

- To re-skill the unemployed so that they may re-enter the work force.
- To provide an incentive for the disadvantaged to bridge the gap between the haves and the have not's in the information society.
- To provide an incentive for those outside the work force to develop computer skills.

The ECDL model as created by the task force consisted of four components – the Skills card, the Driving Licence, the Syllabus and the tests. These were described as follows in 1997 [1] as:

- "The European Computer Skills Card records the progress of the candidate and the dates on which each of the seven tests are completed successfully. When all seven modules are completed the candidate is issued with an ECDL. The modules can be taken in any sequence and the tests can be taken in different test centres and indeed in different countries.
- The European Computer Driving Licence is the full licence and indicates that the holder has satisfactorily completed all seven modules. The document has a similar format throughout Europe. It bears the name "European Computer Driving Licence" in English and in the local language.
- The ECDL Syllabus describes the objectives, content and guidelines for assessment of each of the seven modules of the ECDL."
- The European Question and Test Base (EQTB) defines the questions and tests which candidates must pass in order to attain an ECDL."

Module one was the theory test and modules 2-7 were skills tests. Initially there were 100 theory questions and approximately 20 tasks for each of the practical tests. The time allowed and the pass mark for each of the test modules was defined. The first tests defined the following:

- In module one, candidates had to demonstrate an understanding of the basic concepts of IT, answering six questions selected at random, one from each of six sections. The pass mark was 60%.
- In module two "Using the computer and Managing Files" the test consisted of four exercises and the pass mark was 80%.
- The tests for module three "Word Processing" and module four "Spreadsheets" consisted of basic tasks and advanced tasks. The candidate had to complete one exercise consisting of 8/10 tasks. The pass mark was 100% for the basic tasks and 50% for the more advanced tasks.
- Two exercises were set for module 5 "Databases/Filing Systems", the first required the candidate to set up a small database for a specific purpose, define the structure of the records and enter data; for the second exercise the candidate had to load a database and answer questions about its contents by constructing queries. The pass mark was 80%.
- For module six "Presentation and Drawing", because of different facilities available to the candidate, one question was asked from either a presentation or drawing section
- Module seven was "Information Network Services" and because, in 1996, not all countries had Internet access, two versions of this test were

produced – one a test of skill and one theoretical test. The version taken depended on the facilities available. The pass mark was 80%.

4.1 Quality Assurance

Quality had a high priority and quality assurance procedures were defined and managed by the ECDL Foundation, the licensees, sub-licensees and the test centres. The quality assurance procedures included use of rigorous authorization guidelines and compulsory standards. The tests for ECDL were conducted by authorized test centres. These centres were validated by the licensee in each country using guidelines provided by the ECDL Foundation. The test centres were expected to be operated by course vendors, educational establishments, large organizations or companies. It was the intention that authorized test centres would provide facilities to test not only their own pupils but also persons who wished to take a test without undergoing formal courses.

The results of the tests in Europe were monitored on a statistical basis and any unusual patterns were investigated by the ECDL Foundation. Strict adherence to the syllabus and the use of the standard tests and marking guidelines also ensured that the tests were of an even standard. The ECDL Foundation performed quality assurance audits to ensure that standards were maintained and that the quality of the product was ensured.

Quality is considered as a continuous improvement element within the organization and today, the ECDL Foundation quality statement is as follows [14]:

> *"ECDL Foundation is committed to the development, promotion, and delivery of quality certification programmes so as to enable proficient use of ICT that empowers individuals, organisations and society throughout the world.*
>
> *To meet the needs of all our customers, ECDL Foundation has established a Quality Management System based on the internationally-recognised quality standard ISO 9001: 2000.*
>
> *Adherence to this standard ensures that the processes used by ECDL Foundation to develop and support its certification programmes, are effective, efficient, and subject to continuous evaluation and improvement."*

4.2 Accessibility

In early 2001, accessibility was identified as a major requirement for ECDL. The ECDL Foundation was a partner in an accessibility project, ECDL-PD, which planned to examine ECDL, identify accessibility issues and propose solutions. The outcome of this project [15] was the definition of potential barriers to inclusion within ECDL. At the same time other projects examined accessibility issues; these included projects in the UK, Hungary, Italy, Austria [16] and Greece. The ECDL Foundation set up an international working group, which included representatives of disability groups, to decide on how to address the identified barriers and to collate the results and findings. These were used these to inform future development, procedures and standards [17].

ECDL is committed to accessibility and inclusion. Today, the Accessibility Statement on the web site reads:

"ECDL Foundation is committed to ensuring that no one is excluded from pursuing our certification programmes and actively works to maintain the accessibility of our programmes to all, including people with disabilities. We have been working with our national operators and with disabilities groups to identify and eliminate major barriers to the accessibility of ICT skills and will continue as part of the ongoing enhancement of ECDL Foundation's programmes."

4.3 Development and Updating

The ECDL Foundation set up a Members Forum which gave a platform for licensees, sub-licensees and test centres to share experiences and to suggest developments which they felt were appropriate. This ensured that the products were updated and developed in line with advances in technology and market requirements. The ECDL syllabus and EQTB were updated on a regular basis by an international panel of subject matter experts, ECDL licensees, end users and companies.

4.4 Automation

A priority in 1997 was to ensure that the testing process was automated to the fullest possible extent. This was seen to be vital in the light of the expected volumes of ECDL candidates planned to be some hundreds of thousands initially and ultimately would be millions. The feasibility of taking tests over the Internet from work or home was also be investigated as were diagnostic tests which pointed to the areas of weakness of each candidate rather than the pass/fail testing process [1].

4.5 ECDL Products and Programmes

There was a window of opportunity for the core ECDL at that time in 1997. It was hoped that it would quickly gain acceptance as the test of skills in Europe. There were early optimistic signs that this would be achieved. The support of the European Commission through funding from the ESF (European Social Fund), DGIII and DGXXII gave the ECDL considerable credibility. In addition, ECDL was included in the Information Society Action Plan prepared by Commissioner Bangemann for Central and Eastern European Countries [18].

Countries outside Europe began to take an interest in ECDL in 1998. In these countries ECDL runs as ICDL (International Computer Driving License). The syllabus, procedures, test methods and certification remain exactly the same. The first ICDL was presented to a young woman in Port Elizabeth in South Africa in 1999 [19].

The ECDL which has been discussed in this paper is the core ECDL which is identified today as an ECDL programme – "Essential Computer Skills". ECDL now offers many other programmes. The first to be developed was the advanced ECDL, which was proposed by the Danish computer society in 1999. Programmes developed over the last ten years include e-Citizen and Equalskills – these comprise programmes on "Internet and email" and "computers for beginners", respectively. These programmes fall within the digital literacy categories. Other specialist programmes include Computer Aided Design (CAD), Website creation, Health Informatics and Digital Imaging.

5   The Scope of the ECDL Potential Population 1997 - 2010

Who should benefit from ECDL? The potential users for the ECDL products and services could be seen from several viewpoints, such as geographical, by population sectors, and specifically the work force. The following statistics are taken from the plans presented to the European Commission in 1997 on the setting up of ECDL [20].

5.1 Geographical Scope

The market for ECDL was those countries which were defined as being European by virtue of their membership of the Council of Europe. The Member Societies of CEPIS were present in 17 of those countries in 1996 with a total of some 250,000 computer professionals. The CEPIS Member Societies represented channels to a major part of the overall market.   CEPIS was represented in the following countries: Austria, Cyprus, Denmark, Finland, France, Germany, Greece, Hungary, Ireland, Italy, Netherlands, Norway, Poland, Spain, Sweden, Switzerland, and UK. The potential market was all of the wider Europe. This included the then 15 Member States of the EU, the three Associated States (Norway, Liechtenstein and Iceland), Switzerland, and the then twelve applicant states (Latvia, Estonia, Lithuania, Poland, Hungary, the Czech Republic, Slovakia, Bulgaria, Romania, Slovenia, Cyprus, Malta).   Altogether this added up to 31 countries, which was the number of countries that were addressed in the proposed business and implementation plans for the ECDL at that time.

5.2 Population Sectors

The total population in the 31 countries was about 500 million. Divided in sectors, the numbers were very roughly as follows:

1	the employed work force	200 millions	
2	the education sector	100	
3	the unemployed	25	
4	the socially disadvantaged	15	
5	people outside the work force	160	(not unemployed)

ECDL was relevant to the all users throughout the population. However, it was felt that the Business Plan had to focus on selected parts of the total market. In accordance with the initial objectives, the first ECDL Business Plan primarily addressed the needs and potential in the European workforce.

5.3 The Work Force Sector

The workforce definition varied considerably across the geographical area in question. Very generally, one could have used the following assumptions:

Workforce (employed, 40 % of Total Population)	200 million
Non-manual workers (60 %)	120 million
Knowledge/information workers (50 % of above)	60 million

At the time of the development of ECDL it was believed by the task force that computers were used by some 50% of the workforce. Thus, as an order of magnitude, it was not unreasonable to assume that some 60 million European workers should be competent at least at a basic IT-user level, in order to utilise the investments in information technology and systems. It was a fact that most of these millions of users had become IT-users over very few years and that most of the users had received little or no IT-training in their schooling or education before they entered the work force. At this time, the European workforce probably faced the largest retraining and adult education challenge ever. ECDL addressed exactly this challenge by defining a basic skill level and by offering mechanisms to entice users, employers and organizations to raise the skill level of the individuals in the work force. It was acknowledged that these needs could not be met in the very short term.

5.4 Today

ECDL has grown worldwide with 9.7 million candidates and 121 active countries by November 2009. This growth is shown in Figure 1 below.

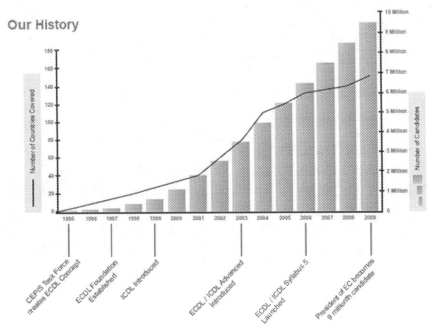

Fig. 1. Timeline and history of ECDL[1]

6 Organisation

The ECDL Foundation was set up in Ireland as a Company limited by guarantee, having no share capital. Following legal advice this model was selected as being the

---

[1] http://www.ecdl.org/publisher/index.jsp?p=94&n=170

most suitable to meet the needs for implementing ECDL throughout Europe. It gave protection to the members, being the Member Societies of CEPIS, as well as allowing the organisation to function as a business with charitable objectives. All surplus revenues were retained within the Company and used to further the objects of the Company as set out in the Memorandum and Articles of Association.

The licensees became members of the ECDL Foundation and are the controlling body. They elect the board of the ECDL Foundation at the Annual General Meeting. The Board of the ECDL Foundation consists of a maximum of seven persons elected at the General Meeting.

It was decided to locate the ECDL Foundation in Dublin because of the infrastructure in that country for supporting the Software Industry. The tax arrangements were potentially attractive and some grants had been secured to help finance the initial set up costs of the ECDL Foundation. The ECDL Foundation needed a very skilled and motivated staff, in order to succeed with the ambitious goals and plans for the organisation. The skills included solid IT, educational and pedagogical backgrounds, combined with marketing, business and administrative experience. Experience with international work and co-operation was also an absolute requirement. As detailed in the first implementation plan, the ambition levels required 4 - 8 staff for the 5-year plan period. It was planned that the ECDL Staff in Dublin would have a central capacity of 3 early in 1997, and then increase to 4. Other people resources would be recruited on a consulting basis from the other ECDL Foundation Member countries, in order to encourage use of Best Practices within the ECDL community. This model ensured flexibility and transfer of knowledge [21].

7 Conclusions

ECDL has gained a reputation as a standard for digital literacy in many countries. The largest adopters of ECDL in total numbers are the UK and Italy. In the latter, all school children undergo training in ECDL. In both countries the Government Departments of Education provided funding in various ways. ECDL also performs well in a social sense being used in prisons in Austria and in Refugee Camps in Gaza. The largest penetration per head of population is in Ireland where over 10% of the working population has achieved ECDL.

ICDL continues to be successfully implemented throughout the world, most recently being adopted in Kenya where the Information & Communications Minister announced that ICDL is to be the sole government recognised entry level computer certification with which to demonstrate competence in computer usage. ICDL has been adopted in Jordan where all teachers must achieve the ECDL and by UNDP as a standard for the training of staff throughout the development programme.

The success of ECDL is due to many factors which have contributed to its acceptance throughout the World. Initially the word "European" in the title helped to give the certification a degree of credibility before it was widely recognised. Also the fact that it was included as one of the 27 projects in "Towards the Information Society – Twenty-seven ideas for European Initiatives" at a time when the European Union was preparing for expansion was extremely helpful [22]. ECDL was developed and disseminated initially through Member Societies of CEPIS which was important in gaining acceptance by Governmental bodies. The ethos and professional standing of the Member Societies added credibility to the certification and ensured that quality and standards were maintained at all times.

In 1995, the project which created ECDL set out to define the knowledge and skills which were needed with the arrival of the new end-user technology. Today, digital literacy is becoming more important and the ECDL programmes continue to develop and address the changing needs for such knowledge and skills.

References

1. User Skills Task Force, Esprit project documents and working papers (1995-1997)
2. Occhini, G.: ECDL the Take off years, Upgrade (4) (August 2009), http://www.cepis-upgrade.org/issues/2009/4/up10-4Occhini.pdf (accessed February 9, 2010)
3. IFIP, The International Federation for Information Processing, http://www.ifip.org/index.php?option=com_content&task=view&id=56&Itemid=110 (accessed February 10, 2009)
4. Council of European Professional Informatics Societies, http://www.cepis.org/index.jsp?&p=636&n=637 (accessed May 10, 2010)
5. Berleur, J., Brunnstein, K.: European Informatics Skills Structure (EISS - CEPIS) code of professional conduct. In: Ethics of Computing: Codes, Spaces For Discussion and Law, pp. 175–176. Eds. Chapman & Hall Ltd., London (1996)
6. Bangemann, M.: Europe and the Global Informations Society (recommendations to the European Council) (1994)
7. Cortada, J.W.: The Rise of the Knowledge Worker. Butterworth-Heinemann, Butterworths (1998), ISBN: 978-0-7506-7058-6
8. Bangemann, Living and working in the Information Society (1996), http://www.uni-mannheim.de/edz/pdf/kom/gruenbuch/kom-1996-0389-en.pdf (accessed February 9, 2010)
9. Bangemann, M.: The Information Society and the Citizen, http://ec.europa.eu/information_society/index_en.htm
10. Gilster, P.: Digital literacy. John Wiley, Chichester (1997), ISBN 0471165204
11. Computer Driving License to the rescue, http://www.e.finland.fi/netcomm/news/showarticle3511.html?intNWSAID=40488 (accessed February 9, 2010)
12. Irish Congress of Trade Unions, open meeting held in Dublin (March 1996)
13. Carpenter, D., Dolan, D., Leahy, D., Sherwood-Smith, M.: ECDL/ICDL: a global computer literacy initiative. In: 16th IFIP Congress (ICEUT200, educational uses of information and communication technologies), Beijing, China (2000)
14. The ECDL Foundation, http://www.ecdl.org/publisher/index.jsp?p=94&n=609 (accessed May 10, 2010)
15. Petz, A., Miesenberger, K.: ECDL– PD — Using a Well Known Standard to Lift Barriers on the Labour Market. Springer, Heidelberg (2002), 978-3-540-43904-2
16. ECDL barrierfrei, http://www.epractice.eu/en/cases/ecdlbf

17. Leahy, D., Dolan, D.: Making an International Certificate Accessible. In: Miesenberger, et al. (eds.) Computers Helping People with Special Needs, 11th edn., Linz, Austria, pp. 1313–1320. Springer, Heidelberg (2008)
18. Public strategies for the Information Society in the member states of the European Union, http://www.ecdl.com.cy/assets/mainmenu/131/docs/EU-MemberStatesStrategies.pdf (accessed May 10, 2010)
19. Dolan, D.: In: Marshall, G., Ruohonen, M. (eds.) Capacity building for IT in education in developing countries. Chapman and Hall, Boca Raton (1998)
20. Presentation to European Commission, Brussels, Internal Working Document, Members of the User Skills Task Force (February 1997)
21. Johnson, R.: ECDL is launched – now what?, Upgrade (2009), http://www.cepis-upgrade.org/
22. Towards an European Information Society Major EU Initiatives, (1996) http://www.euro.ubbcluj.ro/~htodoran/articles/articol1.html (accessed February 9, 2010)

A Brief History of the Pick Environment in Australia

Stasys Lukaitis

School of Business Information Technology RMIT Melbourne Australia
stasys.lukaitis@rmit.edu.au

Abstract. Mainstream Information Technology professionals have misunderstood the Pick environment for many years. The Pick environment has been conceived, designed and built with business solutions as its key driver. At its heyday there were over 3,000 business applications available across a very wide range of hardware platforms supporting from 1 to thousands of real time users. The tentative economic recovery of the 90's and the Y2K fears created cautious and conservative corporate decision-making. During those tumultuous years there were startling leaps in information and communications technology rewarding those who invested in the future and in themselves. The Pick community at the time were fragmented and somewhat narrow-minded in their view of the future and were unable to collectively invest in developing new technologies. Corporate executive peer group pressure to adopt "vanilla" relational technologies and the desire for homogeneity is creating even more pressure on the Pick community.

Keywords: Pick; Universe; Unidata; Prime; Revelation; jBase; Reality; Multivalue; Correlative; D3.

1 Introduction

A simple Google search for the PICK environment or database will reveal a plethora of information about the history of the Pick Operating System and DBMS over the years. It is not the intention of this paper to repeat what is already freely available. Suffice it to say that PICK was 'invented' in the USA by Richard (Dick) Pick and Don Nelson as contractors for TRW on the Cheyenne Helicopter parts and maintenance project. The Appendix shows the various incarnations of the product as well as its early entry into Europe through Microdata Intertechnique in France.

This paper seeks to highlight an interesting historical period in the development of Information and Communications Technology in Australia and to notice the various influences of (then) emerging factors on the Information Systems that were in use at that time, and the Pick environment as an interesting example. The period under investigation is the last two decades of the twentieth century 1980-2000, with the emphasis on the early 1990s and the period approaching the Y2K event.

1.1 Research Approach

Thus this is a hermeneutical analysis of a brief period in Australian IT history using the Pick environment as the driver. Hermeneutics is a philosophy of enquiry that

A. Tatnall (Ed.): HC 2010, IFIP AICT 325, pp. 146–158, 2010.

seeks to gain understanding about an issue or question using techniques that attempt to deal with a researcher's biases and prejudices, and in particular the effects of historicality[1] - not taking into consideration the historical milieu and social events and thinking of the time, and the way that language and its use and interpretations can colour understanding and interpretation [1].

In this usage of hermeneutics I use original documents in the form of books, reports, magazines, articles, quotations from the industry leaders of the time and personal experience. The reason for the use of the hermeneutic philosophy is to glean understanding of the historical milieu from a variety of data sources.

This research will review the "forgotten factors" of the time, the drivers that pushed the Australian IT industry, the key decision makers and what was happening to technology then. Historical investigations such as this should be free from the emotion and biases because, as Gadamer stated, the passage of time has allowed the events to be "closed" [2].

The Pick environment in the first decade of the twenty-first century has become a little more obscure than during its heyday in the latter part of the twentieth century, and this article might shed some light and perhaps some useful debate on the issue. The understanding and appreciation of history in its unsanitised form can be helpful in avoiding mistakes and errors already committed. It is a fact that with "*all histories they are the tales of the winners who always rewrite history to their image, leaving many stories untold*" [3].

2   Brief Overview of the "PICK" Concept

Without experience in the Pick environment it was easy to be confused at the time as to what Pick actually was. Mostly it was believed that Pick was just an operating system with some sort of database management system also called Pick. Operating systems in the 1970s and 1980s were mostly proprietary and non-portable between hardware platforms. As an example, HP minicomputers were released with a proprietary operating system called MPE-IV (1980's) [4] that only worked on HP manufactured equipment. Similar examples existed with IBM and DEC (VMS).Open systems were emerging and one called Unix was starting to be seen available on several different hardware platforms. This portability was exciting the ICT community at the time because being tied to a particular hardware vendor was seen as undesirable for many reasons, one of which was being locked into a scalability ceiling[2]. IBM was famous at that time by offering relatively continuous scalability solutions right through to their mainframe systems. This section briefly outlines the Pick environment from four perspectives of a) the data model; b) the operating system; c) applications development environment; and d) the data retrieval model. Readers interesting a more thorough explanation of this environment are referred to the bibliography for a range of material.

---

[1] Historicality: A term coined by Hans Georg Gadamer to describe the effect of time and cultural distance between the investigator and that being investigated. An example might be the difficulty in understanding life in ancient Rome from the perspective of a 21st century Los Angeles dweller.

[2] Scalability ceiling: Reaching the performance limits of the hardware/operating systems platform and being unable to sensibly expand beyond that capability.

2.1 The Data Model

The actual database was modeled as a "Hashed Indexed Sequential Access Method" (HSAM) mechanism. Each data record was indexed by a unique primary key that is used to calculate which frame (or bucket) is its home location inside a given fixed-size file. So for a given file and frame the read and write commands "compiled" directly to an absolute disk, head and sector address.

Unlike other data models Pick's record structure is not predetermined by a Data Definition Language. Traditionally a database is created with the required number of tables, each table having its own peculiar structure. In Pick, one created files as

Table 1. Student Record Layout in Pick

Attribute #	Data
0..11	Traditional fields student ID, names, addresses, sex, phone numbers, DOB
12	Subject code (text)(MV Controlling)(Associated to file Subjects)
13	Semester code enrolled (text)(MV Dependent #12)
14	Result (number)(MV Dependent #12)
15	Assignment result(s)(MV Dependent #12)(Multi-sub-valued- multiple assignment results allowed)
16...	More attributes

When we look at an actual record, it might be represented like this …

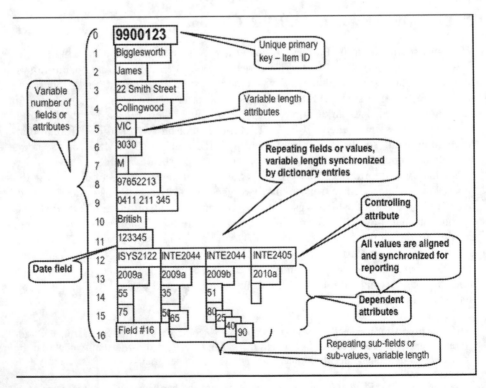

Fig. 1. Representation of a Student Record

needed, each one equivalent to a table. The database was then all the files that were related. Typically one would create an account called "Student Records" and all the related files would be stored there.

Pick differs from other database models because it allows fields to have repeating values and for one field to be a "controlling" field with others defined as "dependent" fields. This allows synchronization of repeating fields. It is also possible for any of these individual repeating fields to themselves store repeating "sub-fields".

Whereas people using the relational model must normalize their structures, the Pick model allows avoiding first normal form and the consequential join tables. Look at the example of a student database that stores student data as well as multiple subjects studied and as well as results for multiple assignments for each subject.

The repeating fields at 12, 13, 14 and 15 break Codd's 1NF law but allow this repeating data to be stored inside the original record instead of creating an additional table that will need to be 'joined back' later when accessed. Attribute 15 holds repeating sub-values aligned to the controlling attribute 12. Here we are able to model assignment results for a given subject where there might be a varying number of assignments for each subject and even varied according to the semester taken.

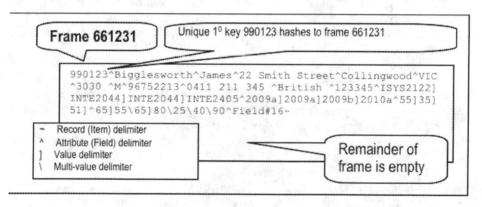

Fig. 2. Example of the Actual Data Record on Disk that is Read and Written

Thus one of the key differentiators of the Pick data model is that each "master record" can also contain all the detail elements associated with transactions on that record. In the example shown, the one structure has replaced several relational tables, or indeed several network-based DBMS detail files. This is quite a saving in not only potential disk space but more importantly in the disk I/O necessary to perform queries and updates.

2.2 The Operating System Model

There was a stranglehold by the major vendors on their proprietary hardware and attached host operating systems. IBM, HP and DEC were but a few of the major players. It is now of historical interest the difficulties that IBM found themselves in with OS360 described by Brooks [5]. The Pick operating system was multi-user and time-sharing with the ability to run dozens of serial users on an Intel 486 computer with

512k RAM and an RS232 expansion card. Pick was an early implementer of code reentrancy which enabled efficient working set management to be implemented [6].

Perhaps from the Pick perspective, two important events were the release of the RISC engine as popularised by the Motorola M68x and IBM RS6x series chipsets and the implementation of SCSI hard disk technology which allowed very fast disk access. Pick was ported to the M68xx chipset and arrived in Australia as the Wicat computer. This was one of many ports of the operating system.

2.3 Applications Development Environment

The applications development environment was implemented with a programming language that was tightly integrated to the host operating system and the database management system. The language syntax included very sophisticated string manipulation capabilities and dynamic arrays that mimicked and implemented the database's fundamental record structure, and internal conversion routines that allowed quite advanced date and time manipulation.

Specialised syntax allowed rapid read/write access to any files that your security level allowed. You could read an individual record or even fields within that record. Locks could be applied on a record to prevent file integrity problems surfacing in a multi-user environment to the degree where an optimistic locking strategy could be programmed to avoid a race condition and ultimate deadlock occurring.

The language compiled to a p-code[3] that was remarkably efficient in a multi-user environment and later into native chipset executable code which increased execution speed markedly. It supported run-time relocation of subroutine code. This was keenly exploited by programmers storing the names of candidate subroutines in files that were read during program execution and loaded and run according to state conditions. Recursion was also supported, although not greatly used as it (still) is a dangerous tool in the hands of any novice programmer.

Most Pick developers were conscious of the power in the programming environment and took appropriate precautions. Nevertheless, as Dick Pick was once quoted as saying that his system was replete with features he called "rope". There was plenty there with which to hang yourself [7].

The only problem with this development environment and language was its name – Pick/BASIC. Most developers in the Pick community thought nothing of its name, but were constantly embroiled in bickering about the language with non-Pick developers who imagined it to be a form of Dartmouth Basic.

2.4 Query and Reporting Language Environment

The query language, called ACCESS, used against a given file would be driven by the data dictionary associated with that file. In other words, using the student record example earlier, you could readily query the student file for reports or information from the target file name (student) and the list of fields (dictionary definitions) in which you were interested. This is consistent with most modern query languages such as SQL. Unlike modern day SQL, Pick ACCESS is strictly a reporting language.

---

[3] P-Code. Pseudo code. A code that was not unlike assembler, but was interpreted at runtime by the internal Pick OS engine, not a p-code compiler.

2.5   Diversity

There were numerous implementations of the model from traditional Pick (OS, programming language and DBMS) through to jBase (Unix, Linux or MS Windows OS, any DBMS and any language but still including Basic). Appendix 1 illustrates the development of the environment in its three streams – the traditional Pick "R83", the Microdata Reality branch and the Prime Information/U2 branch. Today there are several implementations of this Pick model and depending on your scale and communications needs, there will be several implementations from which a good choice could be made.

3   The 1980s and 1990s Context

3.1   Database Choices

In those early days of databases there were a couple of choices of database model. Indexed Sequential Access Model (ISAM) databases were the most popular amongst the larger machines and were used by IBM in their offerings and made "famous" by the leading database writers of the time like James Martin [8]. The other popular database model at the time was the network model, a precursor to the relational model of today. This was popular amongst mini-computer vendors typified by Hewlett Packard and the IMAGE databases [9].

Finally, the iconic Ted Codd published a number of papers describing the two-dimensional relational database model, based on predicate logic and a relational algebra and calculus [10-12]. Numerous implementations of this relational database were spawned with some examples being DB2, Informix, Ingres, Sybase, Unify, Progress, Oracle (various hardware platforms) and later, some Open Source Unix implementations such as PostreSQL and MySQL[4].

3.2   Universities' Impact

Universities around the world became enamoured of the possibilities of a DBMS that was based on a mathematical model. This was understandable as many computer science departments were born from mathematics schools [13].

Database courses were taught in Universities at the time used the popular book by Tsichritzis (Data Models) [14] that identified three main data models – the relational, network and hierarchical models and further treated on the E-R[5], Binary, Semantic and Infological models. These were the only models taught.

3.3   Centralised IT Dissatisfaction

Mainframes were becoming increasingly unpopular because of the hold by DP[6] departments on business applications development (COBOL, PL/I, RPGII, etc) and delays and errors in delivery of systems. Anecdotal evidence at the time estimated the

---

[4] MySQL: Bought out by Sun Microsystems which is now owned by Oracle.
[5] E-R: Entity Relationship data model.
[6] DP: Data Processing departments as they were known at that time.

applications development lead time to be about 4 years. The inability of centralized DP departments to provide satisfactory service levels (sic) led to local "Departmental" solutions leveraging off the emergence of the new mini-computers. Many of these solutions were sourced from enterprising companies who could see the need for bureaus, a service that companies or Departments could buy to solve pressing IT problems.

3.4 Microcomputers and Applications

Visicalc[7] empowered the accounting fraternity with the ability to create complex budgets and perform "what if" scenarios. This released accountants from the control of their DP groups and mainframe-based computer financial models and gave them the independence to plan, model and forecast without the constant DP engagement.

The word processor and "Windows" desktops heralded workstation ubiquity and the death of the beloved typing pool and the start of end-user computing. The "electronic office" was coined and a "paperless office" was promised.

3.5 The Internet

The second thing after PCs and VisiCalc was international data communications, Usenet and email. In hindsight this event created the birth of the global village. By subscribing to your local ISP of the time you could plug into the wisdom of thousands of savants and exchange electronic mail with them and others instantly. Back in the 1980's some quite serious problems could be tackled by joining the appropriate user group and taking a few tentative steps in asking for help. Like today's Wikipedia and Web2.0 the newsgroups were dominated by the loudest and most shrill voices and those with the most apparent authority. It is not surprising then that many vendors spent considerable budget ensuring that their message was being received loud and clear.

3.6 Point and Click Paradigm

In the middle 1990's the web browser was invented to take advantage of something called hypertext and HTML became the way to browse online information. A pleasantly crafted windows environment with mouse-based point and click browsing became the dominant paradigm for interacting with the now ubiquitous desktop PC and MAC. New forms of data entry exploded with radio buttons, pull down menus, check boxes, and of course hyperlinks, all driven by the mouse as the locating tool.

Products such as Visual Basic, "Powerbuilder" and "Oracle Forms" enabled this windows paradigm to be extended into database interaction and transaction processing environments. So it thus became the dominant user interface paradigm – windows, point and click, buttons, boxes and pull-downs, an entire event driven environment. HTML extended it to include hyperlinks.

3.7 Australian SMEs

Because of simple scale factors, Australia had a lot of small to medium enterprises (SMEs) for whom mainframe solutions were inappropriate and who might have

---

[7] And later and more ubiquitous was LOTUS-123.

turned to a bureau solution. The new mini-computers [15-16] became attractive options and were actively pursued. Companies whose IT was agile enough to "move with the times" often gained significant competitive advantage by simply having the better IT solutions of their competitors.

Examples of this were the burgeoning Credit Union movement in Australia, sophisticated Insurance and Library solutions, and very popular manufacturing and distribution systems reminiscent of the original TRW product and precursors to the MRP and MRP-II[8] products.

3.8 The Pick Community in 1992

Australia had several State-based Pick User Groups active in 1992 and most were called the "International Pick Users Association" (IPUA). The Melbourne group was active and hosted several conferences called PickLab from 1988 till 1994. The 1992 conference attracted several overseas dignitaries and featured prominently in the international press [17].

Local and international Pick identities were canvassed in 1992 to offer their opinions on the future of the Pick marketplace. Table summarises their views.

Table 2. Leading Australian Pick Identities Views on Pick's Future

Identity	Company	Views
Peter Fenwick	Fenwick Software	Pick's future is rosy because of its acceptance of Unix (Open Systems) as a host operating system. Pick is far more efficient than its competitors so one can run more users on equivalent hardware [18].
Rob Coulson	Idealogy Systems	Unix as the host operating system is inevitable and needs to be adopted. Pick's success will be driven by its breadth of applications [19].
Terry Leister	VMark Asia Pacific	Unix is the host operating system of the future. Pick needs to accept new technologies such as client-server models and allow processing to be done on the desktop. GUI such as X-Terminals is needed [20].
Tom Couvret	Prime Computer	The key to success will be standards, portability between platforms, communications, distributed databases and desktop integration [21].
Al Dei Maggi	Sequent Computer Systems	Pick's future depends on it ability to integrate with dominant industry standard software layers and systems such as Oracle. Commitment to supporting advanced GUIs and intelligent desktop integration [22].
Charles Cave	Unidata Australia	Adherence and support of Open Systems standards like SQL, TCP/IP X.25 and X11 based GUIs. Integration with products such as Framemaker (publishing), WordPerfect and various graphical applications [23].
Tim Cianchi	Apscore International	Pick cannot survive in its current form. New object oriented technology will sweep everything away in the next 5-10 years. In the short term Open Systems compliance, integration with desktop applications and adoption of three tier architectures [24].
Mike Ferris	UniPix	Survival means interoperability with the likes of Oracle, Informix, Sybase and Ingres where data can be exchanged at a whim. Pick systems will be sold off within 5 years [25].
Frank Gibb	Blue Circle Southern Cement	The Pick operating system will disappear and the DBMS legend will include SQL support and perhaps a port into Latin language and hosting on a games console [26].
John Buchanan	Triad Software	The three prongs of Open Systems – portability, scalability and interoperability. SQL compliance and seamless integration with new products such as an Excel spreadsheet [27].

---

[8] MRP: Materials and Resource Planning. Precursors to modern ERP packages.

Table 2. (*Continued*)

Identity	Company	Views
Barry Churchill	NRMA	Pick must employ the principles of TQM in continually addressing seamless integration with today's (1992) technology. Pick needs to listen to its users to understand what their needs are. After all, it's the "users" who will buy the business systems [28].
Bob Highland	General Automation	Coexistence with LANs such as Novell's IPX. Pick should make sure that it is interoperable with ubiquitous PCs and other workstations [29].
Alan Glassman	BIX	Pick needed to be ported to the 'state-of-the-art' 3270/RJE and HASP protocols. Microsoft to re-introduce command line for Windows/LANMan and Client/Server systems. ADDS terminal division releases a toast-r-oven connection to their 9000 series terminal [30].

The recurring themes from the Pick community were that Pick as an operating system was doomed and that Unix was to be the host of the future. Interoperability with the burgeoning PC marketplace and other systems (RDBMS[9]) was the next theme followed by the emerging industry standards such as SQL. Networking was addressed by some and its relative low profile in these discussions indicated how little the Pick community thought about TCP/IP, LANs and distributed databases.

3.9 Pick's Popularity

During the 1980's through to Y2K Pick boasted that it had more business solutions available than any other environment. There was even a publication called "The Business Software Catalog" that detailed over 3,000 such business applications [31]. The vast library of systems written in Pick/BASIC could be ported from single user machines through to high-end symmetric multi-processors with redundant non-stop capabilities supporting thousands of real-time users.

By this time there were numerous flavours of the Pick environment available. They are detailed in the appendix which illustrates clearly the competition in the Pick community. In addition there were several integrated development environments based on Pick called fourth generation languages (4GL) or CASE[10] tools that ranged from elementary program generators such as Wizard[11] to truly configuration and data driven enterprise-class engines such as Cuebic[12], SB+[13] and Posh[14].

In the 1990's it was not uncommon for an organization to identify a software solution that it needed and to purchase (separately) the hardware platform upon which to mount an operating system (such as Unix), a version of Pick (such as Universe), a 4GL (such as Posh) and the accounting, finance and HR system written in Posh. The average CFO could not understand why it was not possible to get everything from one

---

[9] RDBMS: Relational Database Management System.

[10] CASE: Computer Aided Software Engineering.

[11] Wizard: An early program generator that "created" PICK/Basic code from user entered parameters to describe transaction processes against Pick databases.

[12] Cuebic: A name alluding to the three dimensional nature of the Pick data model.

[13] SB+: System Builder plus.

[14] Posh: An acronym for "Port Out Starboard Home", the preferred window allocation on trans-Atlantic boat trips. The name was adopted by its designer and developer Warren Dickins, a boating enthusiast.

shop on the one invoice, like with an IBM solution. There was great end user flexibility in the choices available and numerous vendors each able to supply the 'perfect solution'. The competition was consequently brisk and in high value cases, immensely pressured. The various 'Pick' vendors robustly vied with each other for the prize of the Operating System licence that was often tied to a hardware platform because of the limited ports made by that vendor.

The other edge of the sword was that when errors or difficulties arose it was often difficult to identify the culprit – hardware, operating system, DBMS, 4GL or applications software and it was not uncommon for each to blame the other.

As the Y2K event approached, the 'mainstream' RDBMS community was mounting a major marketing campaign guaranteeing Y2K compliance and promising many years of trouble-free use with their "best of breed" and "world's best practice" products. This idea appealed to many "C" level executives and a lot of organisations who were struggling with COBOL, RPGII and PL/1-based systems elected to adopt typically SAP or Oracle solutions.

There was comfort in conforming to peer group pressure that was vigorously supported by so many seemingly knowledgeable people in so many forums. After all, if everyone was going that way then surely everyone can't be wrong?

3.10 What Changed?

On Saturday January 1st 2000 the world did not end, electricity kept coming out of power points, water flowed as did gas. The IT community rejoiced in their collective wisdom and ability to prevent the disasters that were predicted. However, a number of important new influences or orthodoxies had emerged and had taken root...

- Alignment between cost and value and quality
- The acceptance of products such as Oracle and SAP and their million dollar price tags by financial controllers seeking Y2K immunity
- How can something costing $150k be possibly as good as something costing $1.5m?
- Microsoft establishes universal acceptability of faulty software
- The now infamous EULA states unequivocally that the software you have bought is not guaranteed to work...
- A new corporate jargon emerges
- Multi-million dollar applications from Europe and the USA with labels such "best of breed", "world's best practice" and "enterprise class"
- Would you be brave enough to argue against a product solution that was "acknowledged the best of breed" for your industry sector?
- Decision making on IT acquisitions went from the IT people to the accountants
- Accountants who have now been liberated from their IT departments are now calling the shots on high value IT decisions
- There is a growing atmosphere of suspicion about the justifiability of the huge Y2K expenditures [32]
- The Windows GUI is the only practical user interface to a computer
- The Internet (and TCP/IP) is now the accepted data communications orthodoxy
- Safety in numbers

- In the older days, nobody was ever sacked for buying IBM
- Corporate leaders meet and compare notes about their respective IT solutions
- Universities control the corporate thinking
- Thousands of graduates have only ever been exposed to the Codd relational model and cannot conceive of other models
- Students are increasingly educated using artificial business models that neatly fit their relational toolkits and avoid real-world complexity

The Pick community was certainly not united. Even the local Victorian IPUA attempted to assuage their vendors' sensibilities by renaming themselves to the "Multi Dimensional Database Forum" to remove the word "Pick" from the association name. Few licencees embraced TCP/IP and fewer still acknowledged SQL and interoperability, and those that did make such an investment ensured that it remained proprietary and certainly not portable to other Pick vendors.

4 Conclusions

The Pick community's inability to work together on keeping up with technology will be one of the reasons that many organisations will be electing to drop Pick as their preferred business platform and select more expensive mainstream solutions.

There were some efforts to create a Windows-based front end but it came to nothing as funding was limited to one vendor. It is also my own opinion that industry was so used to the two dimensional database model that interfaces tended to mirror the spreadsheet in concept. Pick's data model being inherently three (or even four) dimensional was difficult to effectively represent with the tools of the day. It is of passing interest that today's advanced HTML is now capable of such a representation without too much difficulty.

It is also curious to note that no popular operating system has been named after its inventor. Had PickOS been named something like OSMV (OS Multi-value) then a lot of criticism might have been avoided, and many days in the California Superior Court likewise avoided. And had Dick Pick named his programming language similarly after the famous courtesans like Pascal and Ada, perhaps something like "Zion" might have reduced the criticisms of the language name. But it is perception that matters today and perception is reality. Perception is a controlled substance and today the accepted orthodoxies are Windows point and click, Internet, Unix, rigorous Codd relational model and it just has to be "world's best practice", whatever that means. With accountants independently making the key decisions today, businesses are happy to change their business models and processes to align with some "best" standard from Europe or America as manifested in a small selection of "Enterprise" systems. Thanks to the largest software manufacturer in the world, it doesn't necessarily have to work entirely properly either.

References

1. Gadamer, H.-G. (ed.): The Historicity of Understanding. Critical Sociology, Selected Readings, Connerton, P. (ed.). Penguin Books Ltd., Harmondsworth (1976)
2. Gadamer, H.-G.: Truth and Method, 3rd edn. Continuum, London (2004)
3. Forrester, C.: Interview (2010)
4. Hewlett-Packard, MPE-IV Software Pocket Guide. Hewlett Packard, Cupertino (1981)

5. Brooks, F.P.: Mythical Man Month. Addison-Wesley, Reading (1975)
6. Denning, P.J.: The working set model for program behavior. Communications of the ACM 11(5), 323–333 (1968)
7. Rodstein, H.E.: Pick For Professionals - Advanced Methods and Techniques. In: Sisk, J.E. (ed.) The Pick Library. TAB Professional and Reference Books, Blue Ridge Summit (1990)
8. Martin, J.: Computer Database organization, 1st edn. Prentice-Hall Inc., Englewood Cliffs (1975)
9. Volokh, E.: Relational Databases vs IMAGE: What the Fuss is all About, in Interex, Detroit (1986)
10. Codd, E.F.: A relational model of data for large shared data banks. ACM Commun. 13(6), 377–387 (1970)
11. Codd, E.F.: Extending the database relational model to capture more meaning. ACM Trans. Database Syst. 4(4), 397–434 (1979)
12. Codd, E.F.: The relational model for database management: version 2, p. 567. Addison-Wesley Longman Publishing Co., Inc., Amsterdam (1990)
13. RMIT, Advanced College Handbook. RMIT Advanced College, Melbourne (1985)
14. Tsichritzis, D.C., Lochovsky, F.H.: Data Models. Prentice-Hall, Inc., Englewood Cliffs (1982)
15. Brinton, J.B.: New minis push into power era. In: Electronics (1979)
16. Wade, L.: Superminis: Evolution or quantum jump. In: Digital Design (1979)
17. Giobbi, G.: Pick Down Under. News and Review, Los Angeles (1992)
18. Fenwick, P.: New Directions in Information Technology. In: Pick-Up. International Pick Users Association (NSW), Neutral Bay (1992)
19. Coulson, R.: The Future of PIK in Australia. In: Pick-Up. International Pick Users Association (NSW), Neutral Bay (1992)
20. Leister, T.: Where is Pick Going? In: Pick-Up. International Pick Users Association (NSW), Neutral Bay (1992)
21. Couvret, T.: Pick - the Next 10 Years. In: Pick-Up. International Pick Users Association (NSW), Neutral Bay (1992)
22. Maggi, A.D.: Pick - From Proprietary to open Systems. In: Pick-Up. International Pick Users Association (NSW), Neutral Bay (1992)
23. Cave, C.: Where is Pick Heading? In: Pick-Up. International Pick Users Association (NSW), Neutral Bay (1992)
24. Cianchi, T.: Will Pick (and the software business as we know it) Survive? In: Pick-Up. International Pick Users Association (NSW), Neutral Bay (1992)
25. Ferris, M.: Where is Pick Headed in the Next 5-10 Years? In: Pick-Up. International Pick Users Association (NSW), Neutral Bay (1992)
26. Gibb, F.: Pick Beyond 2000 - Back to the Future. In: Pick-Up. International Pick Users Association (NSW), Neutral Bay (1992)
27. Buchanan, J.: Whither Pick or Wither Pick? In: Pick-Up. International Pick Users Association (NSW), Neutral Bay (1992)
28. Churchill, B.: The future of Pick - a User's View. In: Pick-Up. International Pick Users Association (NSW), Neutral Bay (1992)
29. Highland, B.: Whither Pick in the 90's? In: Pick-Up. International Pick Users Association (NSW), Neutral Bay (1992)
30. Glassman, A.: Predictions for 1992. In: Pick-Up. International Pick Users Association (NSW), Neutral Bay (1992)
31. IDBMA, Industry Impact Study - the Pick Marketplace. International Database Management Association Inc., San Diego (1989)
32. Lukaitis, S.: The key issues that impact on the successful alignment of business and its IT function, in Department of Information Systems. Melbourne, Deakin (2010)

6 Appendix Various Incarnations of Pick over the Years

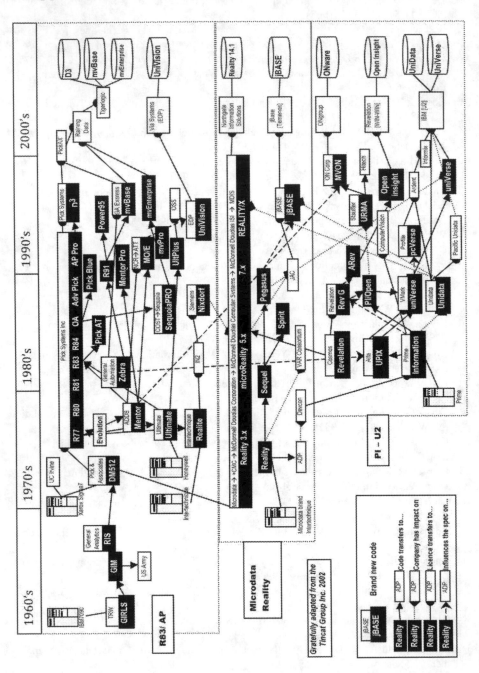

Turning Points in Computer Education

Bill Davey[1] and Kevin R. Parker[2]

[1] RMIT University, Melbourne Australia
bill.davey@rmit.edu.au
[2] College of Business, Idaho State University, Pocatello USA
parkerkr@isu.edu

Abstract. This history of computers in education covers two continents. By analysing the experiences of two people a set of turning points in history is identified. These turning points are in the experienced history and so indicate the impact of changes on the citizens of two countries.

Keywords: Computers, education, history.

1 Introduction

One view of the history of computers and education starts with the use of computers like CSIRAC, used some educational purposes starting in November 1949 [1]. We can be certain that large scale roll outs of educational programs had been established by 1968 as, in that year, reports were being published regarding the success of the use of PLATO terminals [2-4]. Despite the certainty of facts supporting this view of history, questions remain: "How widespread were computers in education?" and "what were the real impacts on people's lives from the computer and when did they become real for the ordinary person?"

Trends in history can be seen by following the careers of giants and pioneers. Another view can be obtained by following the paths of those who are swept along by history. This paper examines the career of two information systems academics from either side of the Pacific, one Australian and one American. These two societies, which were responsible for producing the first computers and using them as educational aids, are considered to be at the forefront of computers in education. A narrative of these two lives shows trends and helps to identify the turning points in the impact of computing from the standpoint of interested bystanders. Rather than laying out the narrative in chronological order we capture the history of the two subjects in terms of the important stages of computers in their lives.

2 Method

To prepare this paper the two authors started with an initial brief: "what was the impact on our lives of information and communication technologies." This resulted in two narratives that were prepared independently, recounting the circumstances that each person thought of as significant. The narratives were then swapped and some

A. Tatnall (Ed.): HC 2010, IFIP AICT 325, pp. 159–168, 2010.

content added, on each side of the Pacific, as a result of memory being stimulated by the other narrative. After some dialogue a set of themes was determined that captured the commonality between the narratives.

3 High School

There is almost a decade in age difference between the two authors. This age difference had its most profound effect during the authors' high school years. Those years, during the 60s and 70s, coincided with an increased availability of computers. Thus, the narratives commence with their high school experiences to show the extent to which the change during that time period affected the average high school student.

The Australian attended high school in the 1960s before high school computing courses became available. Technology affected his life greatly. The launch of Sputnik in 1958 created an atmosphere of panic in the USA and Australia that was reflected in immense changes in the emphasis put on the sciences. American educators became convinced that the Soviets had been able to launch a satellite before the USA because of a superior education system. Intense efforts were devoted to improving the education system, particularly in the sciences, and they spread throughout the Western world. The Australian education system in 1964 saw the introduction of a new experiment-based physics course known as PSSC (Physical Science Study Committee) Physics. This meant that high school students in Australia were introduced to the idea that students could learn by exploring and understanding rather than by repetition. In this environment the Australian, an above average student, was swept into an interest in and love of the physical sciences. The Australian student had to drop subjects seen as not important to a university science course such as Latin, geography, and history. As an above average student he was directed away from commerce courses involving typing, bookkeeping, and office practices into the "important" subjects of English, pure and applied mathematics, physics, and chemistry. Contemporary history shows that computers were being distributed through the public service and some businesses at this time, but there was no mention of them in high schools.

The American attended high school in a small town in the mid-1970s. The local high school offered data processing courses taught by the vocational/technical education department. Vo/Tech, as it was called, also offered classes in automobile repair and other "blue collar" areas. Students who aspired to attend college were advised by school counsellors to steer clear of Vo/Tech classes in order to take "more rigorous" courses. Therefore the American had no computing-related courses in high school. Fortunately his older sister-in-law got a job in data processing at the local junior college, and sometimes brought home a portable remote terminal and a primitive modem to demonstrate simple games or how the computer could generate drawings of the cartoon characters like Snoopy through the placement of alphanumeric characters. The American student was intrigued.

It can be said of both students that computers had not really affected their high school lives and formal study. It can also be said that the environment of their teenage years would predispose them to an interest in computers when they became more available.

4   University

The Australian chose to attend the newly created Monash University in Victoria. At the time this University seemed to be well ahead of its time. They offered a physics course that included coverage of quantum mechanics, relativity, and the type of celestial mechanics related to space travel. This seemed greatly preferable to the more established Ivy League university in which physics still included the study of steam power. By coincidence this University was also committed totally to computerization. Before building had commenced on the University the Interim Council was interested in establishing computer resources at the new university. They believed that once it was operating the University should use the new computing devices for student record and enrolments. An extract from their minutes dated 21 March 1960 notes, "The Council strongly supported proposals that students' records and accounting be mechanized, as far as practicable, from the outset" [5].

This commitment had not extended to the first year classroom by the time our Australian attended classes in 1966. In the first year the statistics classes did involve the use of mechanical calculating machines. The user entered data by pushing keys and pulling a handle to make the machine calculate the answer. By second-year these machines included an electric motor that did away with the onerous task of pulling on the handle. In the third year computer programming classes became available. Behind these classes was a long (for a new University) history of solid work. Australian computer pioneers had brought computers into the University in 1962. The Monash history [5] records the following:

> The 7000 word machine arrived on board ship in Melbourne in early 1962. Dr. Cliff Bellamy, then employed by Ferranti, worked with Professor Westfold and Dr. Sinclair to create a proposal for a joint computer installation at the University. The University would provide space for the temporary loan of the 7000 word Sirius, with two sets of Ferranti/Creed model 75 tape editing equipment in a room fitted with office furniture and telephones.

The Australian started with a class using the Ferranti and punched tape, requiring knowledge of Hollerith code and machine language. By his final year he was using IBM punch card machines to write programs in the user-friendly FORTRAN language. This revolution seemed like a paradise compared with the tedious "one chance" coding onto tape. On reflection the idea of typing cards in the early hours of the morning and having a one or two day turnaround before finding that a card had been placed out of order or a batch card was missing does not seem so pleasant. The point here is that in the time frame of a year or two the computing facilities had changed so much for the better that the student could see a rapid spiral of access that would make computers exciting for some time to come.

The American enrolled in the local junior college in 1977, the same college at which his sister-in-law had once worked. He was taking his general education courses prior to attending a university, and decided to pursue his interest in computers by taking a programming course offered by the data processing staff. The course taught the basics of programming using FORTRAN IV. Programs were written on special forms called coding sheets and then converted to punch cards using a keypunch machines with a typewriter-like keyboard. Each line of code required an individual punch card. Groups or "decks" of cards combined to form programs. Students submitted the program decks, generally followed by data cards to be read by the program, to a person working behind a counter in the computer room. Decks contained special Job Control Language cards to act as job separators so that an operator could stack several job decks in the card reader at the same time and be able to quickly separate the decks manually when he removed them from the stacker. After running a job the computer operator would return the card deck and any hardcopy output to the student or file it for later student pickup. He did not enjoy the course and decided that his interest in computers had been misplaced.

He then transferred to at the University of Texas at Austin in 1979. He started pursuing a degree in Chemical Engineering, but quickly found that he had little interest in that major. A friend talked him into taking a Pascal programming course, and this time the instructor was more experienced and a better communicator and the American thoroughly enjoyed the course and found that he had an affinity for programming and software development. As with his prior experience, punch cards were used for program submission. Unlike the much smaller junior college, however, wait times could be discouraging The university had multiple computer labs in various buildings across campus. Lines at keypunch machines and for deck submission were often so long that the American student would go to the labs at 3 AM to reduce his wait time. The wait time did, however, teach the American the advantages of dedicating a significant effort to the problem-solving phase so that less time could be spent on the implementation phase, an approach that today's students fail to embrace.

The computer science curriculum was well established at that time, and the American took courses in Assembly Language Techniques, Programming Languages, Programming Applications and Practices, Data Structures, Database Management Systems, Numerical Techniques, Artificial Intelligence, and Computer Systems Architecture.

These stories demonstrate the second turning point, in which computers became available to the undergraduate. In both cases they are seen as a vehicle for executing computer programs and each person saw a large jump in ease of use that sparked an interest in becoming adept with the new technology.

5 First Jobs

The Australian's university education was financed through a government scholarship that required him to teach in the school system for three years. By 1970, when he first commenced full time work, the baby boom was well under way in Australia as it was everywhere else in the post war world. The number of new high school teachers was far below the numbers required by the enormous population of high school students. Because the Australian was a new teacher he received a posting at a country high school. The Institute of Technology was short of mathematics teachers in that country town and the young teacher found himself responsible for all the senior mathematics and physics classes, as well as a part-time job teaching engineering mathematics at the Institute. Since this was before an awareness of the importance of balancing work and home life, the new teacher also decided to undertake a postgraduate course in computing at the Institute. By coincidence and enormous luck the local Institute computing degrees were run by the very famous "Westy" Williams, another of the pioneers of Australian computing. This meant that the courses undertaken involved the very most recent computing products. The FORTRAN IV and COBOL languages were featured in the course.

At the same time Monash University had developed a computer system for schools called the MONECS system. This allowed high school students to create programs in FORTRAN using punch cards, post them to Monash, and get a printout explaining the coding errors within a few days. This was very frustrating for teachers and so the local institute was approached for alternatives. A small hand-operated card punch machine was purchased for a mere thousand dollars (about 6 months' salary for a beginning teacher) that allowed cards to be reliably punched and fit into the local Institute computer. This use of computers, to teach programming using punched cards, was a fairly common practice in Australian high schools. It was not until 1981 that a computer-related subject, Computer Science, became an accredited University entrance course [6]. By that time the Australian had been appointed as a curriculum consultant and was involved in managing this course.

The American had a variety of computing-related summer jobs while attending college. Two summers before graduation he worked as an Assistant Operator/Programmer at his former junior college, performing daily computer operations like routine backups, and also evaluated and upgraded various in-house software packages. Prior to his final year of college he worked as Systems Assistant at Conoco, Inc. in his home town modifying and upgrading existing software packages and participating in the analysis and design of new software. Upon graduation in the early 1980s he was hired full time at Conoco as a Programmer and within a year was promoted to Programmer/Analyst. His first job was to participate in the conversion from an IBM 1800 to a DEC PDP 11/34, including the design of replacement software. The programming language used was FORTRAN 77. It was during that conversion that he learned the importance of thorough documentation for maintainability and modifiability. He was responsible for various development projects including production control software, system interface packages, and report generation packages. As one of two on-site programmer/analysts he performed half of all systems development activities including software analysis, design, implementation, testing, and maintenance.

Thus, the second turning point was the recognition that computers were available more widely than in a University as a massive investment. They could be used to teach in high school as well as to enhance business efficiency. Computers were still seen from a programming perspective, but that perspective had shifted to programming with a purpose, not just as a kind of mathematical puzzle.

6 Microcomputers

The next stage in life is not characterized by normal progression but by the intrusion of a new technology. The first Intel-based microcomputer was released in November 1971 with the first Apple microcomputer arriving in 1976.

By 1975 the Australian teacher was a new father and had been selected as a curriculum consultant in mathematics. This entailed teaching part-time and travelling to other schools part-time to talk to teachers about the latest advances in mathematics education. Because of his history of teaching computing in schools the teacher was encouraged to apply for a grant to obtain one of the new microcomputers coming out of America from the Apple Corporation. The grant application was successful and, in his capacity as a member of the secondary math committee, the teacher travelled around Victoria demonstrating how the computer could be used in schools. By 1980 the school in which he was teaching had installed two microcomputer labs, a university entrance subject was available in computer science, and the whole idea of a two-day gap between writing a program and seeing the results was difficult to imagine. By 1989 when the Australian moved to the university sector the prevailing computing equipment was dumb terminals attached to a mainframe computer. Within two years of commencing that appointment almost all of the undergraduate computing was done on microcomputers. By 1991 the degree program in which he taught was called "Information Systems" and involved some programming, but included also the new disciplines of systems analysis and design, usability, and web delivery. The computer was no longer studied for its own sake, but as a tool for business purposes.

The American took a similar path. While working as a programmer/analyst in early 1984, two local higher educational institutions, his old junior college and a local private university, enlisted his aid to teach programming courses at night. At one institution he taught Pascal on a DEC minicomputer, and at the other taught Pascal on Apple microcomputers. He did this for a year, and then he accepted a full time position at the junior college as a computer science instructor. He was also placed in charge of the relatively new microcomputer lab, which had a conglomeration of IBM, Apple, Radio Shack, and other brands of microcomputers. In addition to teaching the usual courses dealing with fundamental computer science and information systems concepts, programming languages and techniques, and advanced data structures, he also taught microcomputer operating systems and usage, and developed the first courses in using spreadsheets (VisiCalc, then Lotus 1-2-3), word processing (WordStar, followed by WordPerfect), and databases (various versions of dBase by Ashton-Tate).

Less than four years later he then decided that he needed to pursue an advanced degree, and returned to college for a Master's degree in Computer Science (1991) and a doctorate in Information Systems (1995). He has been teaching Information Systems at the university level since graduation.

The microcomputer enabled interactive computing for everyone. Interactive languages and development environments were first seen as a better place for people to learn programming, but rapidly a wider view of the computer emerged. Information Systems had become more central than the hardware itself. Microcomputer applications like word processing, spreadsheet, and database software became essential components of most businesses. Both narratives see this change as being due to the sudden impact of personal computers with their enormous power/price advantages over time share systems.

7   Using a Computer

The final turning point grows out of the ubiquity of the microprocessor and the creation of the Internet. One could claim that the Internet started in September 1969 when the first host computer was connected at UCLA. It could also be argued that the historically significant event was the creation of ARPANET at the end of 1969 or the first email application in 1972 or the transition to TCP/IP on January 1, 1983 [7]. Our measure of significance is the point at which the Internet became readily available to our two boys.

The Australian first found telecommunications useful for completing a graduate diploma in computing by external studies. Using an acoustic coupler and Kermit software on his Apple II the degree was completed late at night from this office in 1982. USENET had been a fascination since completing a postgraduate education degree in 1978, where the thesis was written on a DEC10 using the QEDIT program written by Queensland University. The facilities showed him the potential of the global network, but it was only when a microcomputer was available at home, with email and USE-NET, that it became a daily tool for communication. Finally the modern era of asynchronous communications with everyone in the world was ushered in as he took a leading role in the first Pan Pacific computing conference in 1985. This involved academics and others from around the world communicating exclusively by email.

For the American, widespread use of the Internet and World Wide Web started during his final year of graduate school (1995) and intensified during his subsequent university appointment. His early exposure involved protocols like telnet, Kermit, and Gopher, and search tools like Veronica. After graduating he started using Mosaic to access the World Wide Web, but quickly switched to the "new" browser, Netscape Navigator. His first e-mail client was Pine, but he switched to Eudora. Because of his ongoing affiliation with educational institutions he generally had free Internet access and never had to use any of the online services like CompuServe or America Online. At that time the popular search engines included Yahoo!, Infoseek, Excite, and Hotbot. His first capitulation to using the Windows operating system came with the advent of Windows for Workgroups. He developed his first commercial web site in 1998.

8   A Timeline (Results)

The twin narratives can be summarised by the headings we have used here, but a broader picture is obtained by comparing the individual histories with recognised events during the timeframe. In Table 1 below "recognised" historical events have been selected from the Computer History Museum timeline [8]. These show that events early in the history of computing did not always enter the consciousness of the ordinary person, but later events occurred only a short time before impacting the lives of the two subjects.

Table 1. Selected highlights from the narrative compared with highlights from the Computer History Museum

Year	Selected highlights	Aussie and American Highlights
1941	Z3 built built by Konrad Zuse	
1944	Harvard Mark 1 and Colossus	

Table 1. (*Continued*)

1945	Von Neuman report on EDVAC, first BUG reported by Grace hopper	
1946	ENIAC built, inspiring EDSAC, BINAC, IAS, AVIDAC	
1947	Transistor tested	
1949	EDSAC built	
1951	UNIVAC in USA, LEO in UK, first commercial computers	
1952	Grace Hopper first compiler, IBM releases IBM726 magnetic storage	
1955	First transistorised computer, TRADIC, at Bell Labs	
1957	FORTRAN, DEC corporation and Sperry Rand created	
1960	LISP and Quicksort released	Aussie starts High School - no computing.
1961	IBM has 81% of computer market	
1966	HP 2115 released with BASIC, ALGOL and FORTRAN	Aussie starts University - FORTRAN Programming
1969	RS232 protocol and UNIX released	
1970	Xerox PARC opened, ARPANET created	Aussie teaches FORTRAN in High School using punched cards in first year of teaching. Enrols in graduate computer course
1971	First Intel microprocessor, first email	
1972	8008 processor made	
1973	Ethernet released and first personal computer not in a kit (Micral)	
1974	Mouse invented by Xerox	
1975	Bill Gates and Paul Allen license BASIC as the language for the Altair 8800	American starts High School, DP classes seen as "shop"

Table 1. (*Continued*)

1977	TRS 80, Commodore PET, Apple II, Atari released	American in Junior College. Hated FORTRAN IV Classes
1978	Vax 11/780 and 51/4 inch floppies released	Aussie obtains Apple II for demonstration
1979	USENET, Visicalc and Motorola 68000 released	American changes to U Texas and loves Pascal, pursues computer science degree
1980	First hard disk for personal computers from Seagate	Aussie buys Apple II for home, first PC lab of Apple II in school
1981	First IBM PC with MS-DOS, Osborne portable computer 3.5 inch disks	
1982	Commodore 64, Lotus 123	Aussie uses USENET and email; American graduates and works for Conoco as FORTRAN 77 programmer/analyst
1983	TCP/IP, Apple Lisa, Compac PC clone	
1984	Apple Mac, IBM PC AT	Aussie school changes to IBM PC, American part time teaching
1985		American becomes full time college computer science instructor. Lab of PCs
1986		Aussie organises conferences in computing using email and USENET
1991	Linux	American completes Masters in Comp Sci
1993	Mosaic web browser released for UNIX, Mac, Windows, Amiga versions in 1994	
1994	Yahoo founded	
1995		American completes PhD in Information Systems. First use of WWW

9 Conclusion

By examining the critical points in the life history of these two people we see something of the way in which computing history is reflected through the lives of the ordinary person.

The rise of the computer took many years. Because computers were initially available only to governments as defence equipment, both the Australian and American had limited or no exposure to computers in high school. While the Australian experienced the introduction and spread of computers as a university undergraduate student, the American arrived on the scene at a time when university computers were relatively commonplace. In their first jobs they saw the proliferation of computers beyond the academic world, and with the arrival of microcomputers witnessed the eventual ubiquity of computers throughout society. Many of these advances in computing were invisible to the ordinary person.

Their experience shows that exposure to computers often begins in our educational system. The resultant familiarity with computers and software is then propagated beyond the classroom and eventually extends to the population in general. These dual narratives reveal three turning points:

- from ignorance of the computer to seeing it as a vehicle for executing programs
- from an interesting curiosity to a readily available tool
- from a tool to being ubiquitous device designed to support complex software.

The advent of sophisticated smart phones may mark the convergence between the wireless handheld and the PC Markets, changing the landscape of computing yet again [9]. While computers have evolved from being unattainable, to being a curiosity in the realm of hobbyists, to being commonplace and mundane, to the verge of becoming passé, the field of computing will continue to thrive as newer types of hardware and software are required to support this age of enhanced connectivity.

References

1. Doornbusch, P.: Computer Sound Synthesis in 1951: The Music of CSIRAC. Computer Music Journal 28, 10–25 (2004)
2. Blitzer, D.L., Skaperdas, D.: Plato IV - An economically viable large-scale computer-based education system. In: National Electronics Conference (1968)
3. Smith, S., Sherwood, B.: Educational uses of the PLATO computer system. Science 192, 344–352 (1976)
4. Plato Learning: History of Plato Learning, vol. 2004. Plato Learning Inc. (2004)
5. Monash University: Monash University's First Computer (2009)
6. Tatnall, A.: Curriculum Cycles in the History of Information Systems in Australia. Heidelberg Press, Melbourne (2006)
7. Leiner, B.M., Cerf, V.G., Clark, D.D., Kahn, R.E., Kleinrock, L., Lynch, D.C., Postel, J., Roberts, L.G., Wolff, S.: A Brief History of the Internet. ACM SIGCOMM Computer Communication Review 39, 22–31 (2009)
8. Computer History Museum, California, vol. 2010 (2010)
9. Knowledge@Wharton. As Smartphones Proliferate, Will One Company Emerge as the Clear Market Winner? (2009),
   http://www.whartonsp.com/articles/article.aspx?p=1352785

Existence Precedes Essence - Meaning of the Stored-Program Concept

Allan Olley

IHPST, University of Toronto
91 Charles St., Toronto, ON, Canada
allan.olley@utoronto.ca

Abstract. The emergence of electronic stored-program computers in contain the 1940s marks a break with past developments in machine calculation. Drawing on the work of various historians, I attempt to define the essence of that break and therefore of the modern computer. I conclude that the generally used distinction between computers and precursor machines in terms of the stored-program concept and von Neumann architecture rests not only on differences in hardware but also in the programming and use of machines. Next I discuss the derived definition in terms of machines from the 1940s and 50s to elucidate the definition's implications for the history of computing.

Keywords: Stored-program, von Neumann architecture, computer history, computer architecture, history of software, IBM SSEC.

1 Introduction

The July 1939 issue of Astounding Science-Fiction included an article entitled "Tools for Brains" by one Leo Vernon. This article included a brief history of calculating machines, including mention of Charles Babbage and short discussion of recent scientific uses of punched card machines at Columbia and differential analyzers at MIT. He concludes by speculating on future machines sketching what a "dream machine" for such scientific computation might look like:

> "The dream machine may fill an entire building. It will be operated from a central control room made up entirely of switchboard panels, operated by trained mathematicians, and an automatic printer giving back the results. A physicist has spent months getting a problem ready for solution. He has long tables of numbers, experimental data. These numbers have to be combined and recombined with still more numbers, producing hundreds of thousands or millions of numbers..." [1]

Vernon imagined the subsequent calculation occurring in minutes and with the aid of mathematical tables punched onto paper tape. [1]

Vernon's predictions seem in some aspects prescient and in some aspects completely wrong. Computing machines soon did become very large and very fast, but soon Vernon's dreams of a machine operated by a room full of switchboards would

A. Tatnall (Ed.): HC 2010, IFIP AICT 325, pp. 169–178, 2010.

prove more a nightmare and mathematical tables would fade from use. The point of this paper is to illuminate the source of such predictions. Vernon had just finished reviewing developments and practice in scientific computing up to 1939. In part his prediction simply projected an extension of previous practice. Punched card machines used switchboards and tables punched into machine readable forms. Therefore so would the machines of the future. The developments during and immediately after World War II changed, or added to, the nature of machine computation in fundamental ways.

Many attempts have been made to capture the essence of the modern computer. The goals implicit in such definitions have been various, including attempts to characterize a "computer age" distinct from what came before it and the less reputable task of establishing priority for inventors. A requirement given by most analysts for a modern computer is that it be electronic. Electronic speeds made computers capable of thousands of computations in a second, far outstripping the speed of any human effort. However, Vernon's description of the dream machine implies a machine with just such prodigious speeds. So it is not speed alone that separates Vernon's dream from later developments. In parallel most commentaries suggest that the speed of the modern computer is only part of the story. Accounts point to the design elements that make a modern computer a general purpose machine applicable to a wide range of problems.

An example of these developments in machine computation is the ENIAC. The ENIAC, Electronic Numerical Integrator and Computer, began operation in 1946 and resembles Vernon's dream machine. The ENIAC's electronic tubes allowed computation at speeds a thousand times faster than previous machines and operators controlled the ENIAC by rewiring switch boards. However, the physical rewiring of the ENIAC proved onerous and a new system was implemented in 1948. This system controlled the ENIAC's operations using a series of instructions encoded in a numerical format, a program. The operators preferred the new system of set-up despite a significant reduction in the speed of actual computation. [2]

Traditionally historians attribute the articulation of key design elements of the modern computer to the "Draft report on the EDVAC." The report became synonymous with John von Neumann, but was based on the work of a team at the Moore school, responsible for constructing the ENIAC, lead by John Mauchly and Presper Eckert. The EDVAC was to be the successor to the ENIAC and an improvement upon it. The report details various broad design features of the new machine. The report was circulated among many other researchers who incorporated key design features into their own projects. Examples of seminal features of the EDVAC design were serial operation of the machine, separate mechanisms for calculation and storage, a hierarchical arrangement of memories with a small amount of expensive fast storage and progressively more cheap slow storage and the use of a stored-program for instructions. Often the innovations of the Draft Report are called in summary "the stored-program concept," after a single feature. The stored-program has been summarized as the storage of instructions and numeric data in the same format in the memory of an automatic calculator. [3] This feature makes it possible to perform a conditional branch by manipulating the instructions based on intermediate results. This in turn makes a machine Turing complete, that is with the addition of unlimited storage the machine would be a machine capable of carrying out any possible computation.

There are two things to note about this reduction. First that the stored- program is not the only way to achieve a Turing complete machine, such a machine could be achieved by a mechanism that stored data and instructions in completely different formats. Most obviously a machine with hardware implemented conditional branch and unlimited potential for rewritable storage of separate instructions and data, such as magnetic tapes could be Turing complete. The most striking example of the potential to achieve a universal machine by other means was Raùl Rojas' demonstration that German computer pioneer Konrad Zuse's Z3 machine could in principle perform any finite computation a Turing machine could, if it were given an infinite tape and indefinite intermediate storage for data. This is despite the Z3's incredibly limited repertoire of capabilities. The Z3 could perform basic arithmetic as directed by a long tape of instructions. It was not however a stored-program, rather the program was fixed with a small rewritable storage for numerical data. Also the Z3 could not perform a conditional branch, except to stop under certain conditions such as a divide by zero error. Rojas demonstrated that a branch can be simulated on such a machine by carrying out both paths of the branch and negating, via multiplication by zero, any contributions from the path that would not be taken by a genuine branching machine. [4]

The second thing to note is that as narrowly defined a stored-program machine need not have any resemblance to the machine described in the "Draft Report." A machine with the basic stored-program might have no hierarchy of memory possessing only one form of storage, such as say a rewritable tape, or might store and calculate with the same mechanism in a massively parallel system. Despite this, "von Neumann architecture" and "stored-program computer" are used interchangeably, for example Wikipedia has a single entry for both terms. [5] The equation of these two terms strips them of some important connotations. For example, the problem of delivering new instructions and data to the processing unit of a computer from its memory has become known as the "von Neumann bottleneck", a clear reference to the features of the design from the "Draft Report" rather than a feature of a narrowly defined stored-program machine. [3]

The difficulty of defining the essential characteristics of the computer is already widely noted in the literature. The 2000 anthology of history papers, The First Computers, embraces the ambiguity of the concept and contents itself with discussing a broad range of machines in the 1930s, 40s and early 1950s as the first computers, but not accepting any one of these as the singular first computer. [6] My aim in this essay is to expand on this point detailing some non-material aspects that separate computers from other calculating machines. While certain hardware characteristics may be necessary to create a machine with the capabilities of the modern computer, they alone do not constitute the defining essence of the machine. Rather the practices, applications and expectations that surround a machine play a key role in defining what it is.

2 The IBM SSEC a Stored-Program Computer?

Part of my inspiration for this paper has been my research on astronomer Wallace J. Eckert (no relation to Presper Eckert) and his work at Columbia University and IBM. At IBM, Eckert was associated with the development of only a few machines. One of these was the Selective Sequence Electronic Calculator (SSEC) dedicated in January of 1948. This machine sits right at the cusp of the era of the modern computer, but receives limited attention from histories. The SSEC stored instructions and data in the same format, could automatically manipulate instructions and used this feature to vary operation in response to intermediate results. Therefore the SSEC embodied the narrow definition of a stored program computer. Yet this feature often went unmentioned even in accounts that emphasized the narrow definition of the stored-program computer as a, or the, key element of the modern computer. Later machines would be cited as the first stored program computer. [7]

In light of the emphasis placed on the stored-program concept by the literature, I found it incongruous that this feature of the SSEC would be so often ignored. This led me to pay attention to those accounts that recognized the novel features of the SSEC, but gave reasons to deny it the status of a modern computer. The accounts of the SSEC by IBM historians such as Charles Bashe and Emerson Pugh suggested some reasons. The SSEC's use of relays for the bulk of its "fast" memory meant that its operating speeds were somewhat limited compared to later machines and even the earlier ENIAC, despite its use of electronic arithmetic to perform computations. Perhaps more importantly the SSEC had limited impact on later developments. IBM's first commercial computer the IBM 701 was based on von Neumann's computer at the Institute for Advanced Study. The only machine design to take direct inspiration from the SSEC machine was the IBM 650. [8]

Computer scientist and historian Brian Randell offered explicit justification for his placement of the SSEC. While noting that the SSEC was probably the first machine capable of arithmetic manipulation of stored instructions, what I call the narrow stored program concept, he denied it status as the first stored program computer. He cited its small electronic storage size and claimed it operated more in the style of a fixed program tape machine. In disqualifying the SSEC Randell in part makes reference to the need for electronic speed, since he allows that the Manchester SSEM test machine was the first computer, although not a practical one, despite the fact that its memory was only larger than the SSEC's with reference to the electronic storage and smaller when relay storage was included. [9]

However, it is clear that for Randell the stored program means more than the narrow concept and electronic speed. In introducing the stored program concept he noted its significance as allowing universal computation but added that its lasting significance consisted in the concept of using the computer to help prepare its own programs. Randell saw in the stored-program concept the seed of things like compilers and operating systems. Fixed program tape machines do not realize this sort of flexibility and the SSEC would have had trouble holding its full program in its memory. [9] Elsewhere Randell rejected the notion that merely storing instructions and data in the same part of the machine is sufficient to define the stored program concept, referring to the stored program as a practical engineering realization of Turing's universal calculator. [10] Randell's analysis argues that the stored-program was not merely hardware, but a use and attitude towards the potential of that hardware.

Randell was not alone in viewing the stored program concept as richer than the bare storage of instructions and data together. The late Australian historian of computers, Allan G. Bromley, broke the origin of the "stored program concept" into 10 distinct stages of development. For Bromley the development began with the separate efforts to create digital electronic arithmetic and storage circuits and ended with the development of the idea of hierarchical ideas of machine programming, such as microprogramming with so-called firmware. Bromley saw the development of micro-programming by Maurice Wilkes in 1952 as the end of the development of the stored program concept. The storage of data and instructions in the same format occurs in the middles stages of Bromley's story, distinct from other developments, such as using separate parts of the machine for storage and processing and the development of compilers. [11] Like Randell, Bromley saw the stored program concept as more than mere hardware and more as an attitude towards machine use and design.

Now we come to the other inspiration of my paper: a discussion of the SSEC by French historian Georges Ifrah. In his Universal History of Computing, Ifrah recognizes the SSEC's stored-program nature, however he declares it to be the first "near-computer" in history. Ifrah gives two reasons to deny it full computer status. First its use of mixed electronic and relay technology that he argues was an anachronism and more importantly that the SSEC embodies "inconsistencies of logic." Ifrah argues that the designers of the SSEC lacked a clear theoretical understanding and vision for its operation and so made it unnecessarily complex. Ifrah gives von Neumann and the draft report credit for synthesizing the elements of the computer into a complete theory, while maintaining that in the history of technology there is neither a singular first inventor nor invention but a continuous series. [12]

Ifrah's suggestion that the SSEC design was unnecessarily complex has at least some justification. The machine used 12 500 expensive and unreliable vacuum tubes, whereas later machines would obtain more functionality with far fewer tubes. [7][8] Operators of the SSEC have had mixed opinions. John Backus, father of FORTRAN and an early SSEC programmer would later state that "I think it's an extreme stretch to consider it the first 'stored program' computer." [13] However A. Wayne Brooke, the IBM engineer responsible for maintaining the SSEC, felt that quite the contrary the SSEC did everything a stored-program machine did just in different ways. [14]

Brooke spent a great deal of time in the early 1980s drafting a description of the SSEC and arguing for its status as first stored-program computer. The drafts of this unpublished account reveal many interesting aspects to the machine, including the creation of a special modification to facilitate conditional branch. The SSEC did not have a specific operation for branching and worked instead by arithmetic manipulation of the instructions in the memory. The modification Brooke described allowed the branch to a subroutine to be carried out in a more automatic and straightforward manner. [15] This suggests that the concepts and means of control of the SSEC were in flux after its completion.

Another aspect of the SSEC that suggests its "near computer" status is that it lacked single fixed physical configuration and total control of capabilities by the program. Certain aspects such as access to relays, tape drives and the like were determined by plugging. [8] While relatively minor compared to the extensive rewiring required in the original ENIAC, it suggests the lack of commitment to program control.

3   Essence of the Stored Program Concept

While I remain somewhat skeptical of Ifrah's broad strokes analysis I found its boldness inspiring. In honour of Ifrah's suggestion I take the title of this paper from a maxim of French philosopher, Jean-Paul Sartre. "Existence precedes essence." In this Sartre wished to indicate that the meaning of human life is not given but created by individual human deliberation and action. [16] In a parallel I want to agree with Ifrah that the mere existence of the necessary hardware does not confer the essence of the computer, rather the operators' use of the machine determines its status.

I propose that while the traditional elements of the von Neumann architecture (a large fast electronic memory separate from a processor, serial processing of instructions, a stored program and so on) were necessary for the modern computer they were not sufficient. What was required beyond these hardware elements was a regulating ideal that the machine should be as general purpose as possible and a realization that the instructions themselves were subject to manipulation and generation by the machine. If endless plugging of wires on the ENIAC had simply been replaced by endless hand coding of data and instruction numbers in later machines, then the improvement in power and flexibility of those machines would have been limited. Finally, a stored-program machine should have practical achievements that exemplify these characteristics and aims, not mere theoretical hardware capabilities.

It is illustrative to note that the concept of subroutine libraries were quickly implemented in the first stored-program computers developed. The concept was first proposed by von Neumann and Goldstine working on the EDVAC project. The EDSAC, completed and running in 1949 at Cambridge, was the first machine to employ such a library. The developers hoped to avoid error by using computer code known to be reliable and save time in writing code. They disseminated this technique by publishing a programming manual in 1951 that detailed many of these subroutines. [7] The Manchester Mark I, which became operational a little later in 1949, also made use of a subroutine library. This library was an important resource for those who bought the Ferranti Mark I computer based on the Manchester machine such as the University of Toronto. [17] Note that these subroutine libraries, while only directly applicable to identical machines, served as a means of distributing standard programming techniques. Rather than each new computer having to be operated in a completely new way, techniques could be accumulated.

The reuse of code exemplifies both the standardization that is a hallmark of the modern computer and the use of the machine to make programming easier. In this case the operators take advantage of the fact that computer instructions were encoded in machine readable and therefore machine reproducible form. This feature was hardly exclusive to the modern computer, but is still an important element of its success. The importance of digital copying is even more prominent today as more and more media is distributed digitally in various ways.

Actual preparation of code by machine methods was slower in coming. One of the first programs written at Cambridge was called "Initial Orders." This program converted programs written in programmer's notation of the machine code into the actual binary code of the machine. [7] While a small step, such a scheme represented an understanding that the computer could aid in construction of its own programs. However, operators would have to wait until 1952 before development of more complex programming aids like programming languages began development. [3] Of course it was only in 1952 that computers began to become more widely produced. Therefore developments of these elements of the stored-program concept can be seen to occur relatively quickly. Programming languages also allowed programs to be distributed

relatively directly across platforms and so served to unify machines that were still diverse in terms of hardware.
.

The importance of these elements can again be seen by comparison with the limited development of the SSEC. The SSEC made use of subroutines in its programs often storing them as actual loops of paper tapes in an extensive number of tape drives. However, engineer A. W. Brooke could recall no instance where subroutines were re-used in different programs in the four years the SSEC was in use. [15] The SSEC was never used to generate or check coding because of scarce machine time, but the IBM 604, a punched card calculator, was used to perform simple arithmetic checks and manipulations of the instructions. [18] Finally, just as the hardware of the SSEC left few descendants, the programming techniques also apparently had limited impact. The lack of standardized, or widely communicated, programming techniques in the IBM SSEC show how it failed to achieve the full potential of the stored-program. At the same time the lack of such success helps explain its rapid fall into relative obscurity.

4 The Protean Computer?

Historian Michael Mahoney made a similar claim about the nature of the computer when he suggested that the computer was a "protean" machine, its nature determined by the tasks set and programs written for it. According to Mahoney it has no nature to guide the way it is adopted as compared to other technologies. [19] Whether his position is different from mine or not, I would make a difference in emphasis on certain points. First Mahoney seemed to seat the protean nature of the machine in its mere hardware, but, as I explained above, hardware is not sufficient to achieve the requisite degree of generality. The computer finds wide application because people are ready to use it in that way. Perhaps more importantly it seems computers become ever more universal because that is one of the ambitions of their developers.

Components that would make a computer's features too specific were eschewed by developers, ignored by users or viewed as troublesome, as in the problems associated with optimal coding of magnetic drum memories. In drum memories information was only available at one point in rotation, therefore the computer might have to wait for an entire revolution to retrieve a piece of data or instruction. Users often sought to optimize these retrievals by distributing instructions and data so that minimal time was lost to waiting for the drum to rotate. [3] Such optimization was completely hardware specific though. Later memory technologies became more truly random access and generic.

Also, I wholeheartedly agree with Mahoney's emphasis on seeing how various users and communities adapt the computer to their use. However, this does not mean that the computer is inert. In fact, if the computer were truly inert then adoption would be a straight forward matter. The problem already mentioned of the "von Neumann bottleneck" illustrates one example of the difficulty in reshaping the stored-program computer. Adopting new calculating technology often leads to change in practices and goals. The fading away of mathematical tables as a resource in scientific computation provides a good example.

The SSEC was provided with extensive fast table look-up facilities, including dozens of potential tape drives for tables in paper tape form. This is understandable as Wallace Eckert who oversaw its design had a long history of using tables in his work in astronomy. He had pioneered the use of punched card machines in scientific computation at Columbia in the 1930s and made use of mathematical tables in punched card form. Eckert had chosen as the first problem for the SSEC a computation of some positions of the Moon. The position was given by the summation of a long series of trigonometric functions. In the original incarnation of the program high precision values of the sine and cosine function had been obtained from a table punched onto tape. However, the program for this calculation was later reworked and the high precision values of sine and cosine found via calculation of an arithmetic series. [20] With the rise of the electronic computer extensive mathematical tables would fade from use in complex scientific computation, computers could use such tables, but special provision was rarely made for their use as in the SSEC.

Mathematical tables had been a vital component of computation in subjects like mathematics where extended calculation was done. The investment in these methods existed not only in physical copies of the tables, but in the extensive methods of tabular interpolation developed by practitioners. [21] As mentioned at the beginning of this paper Leo Vernon had predicted in 1939 that the "dream machine" would be able to read in tables with lightning speed. Yet the use of tables or not was dictated by the economies of speed in calculation versus reading off values from storage. Electromechanical punched card machines could read faster than they could carry out arithmetic, but not so early electronic machines.

Also, while the computer's nature may be protean, individual computers were not. Mahoney acknowledged this when he gives a definition of the history of computing as "the history of what people wanted computers to do and how people designed computers to do it." [19] However, it bears emphasis that individual computers were designed to be better suited to some tasks over others and so were not protean. For example, in the 1950s IBM marketed some of its computers as scientific machines and others as business machines. Thus the first IBM computer, the IBM 701, was designated for science, while the IBM 702 was designated a business machine. As a result the IBM 701 used a pure binary representation to store and manipulate values, while the IBM 702 used binary coded decimal and this trend continued for several later iterations of machines at IBM. [8]

The distinction between a binary and decimal machine may seem of little import, since it is trivial to convert input and output. However, many would-be operators who had to program in machine code did not see it that way. Decimal was seen as more congenial to the less technically learned business users. Coding in binary seems to have been a worry even for some scientists. For example astronomer Paul Herget did his own programming on the decimal Naval Ordnance Research Calculator (NORC), but was unwilling to do the same on the binary IBM 704. He suggested the engineers have their pay made out in binary until they saw the error of their ways. [22] In principle a binary and decimal machine might not be much different because after all the stored-program meant they could perform all the same calculations. Again human practice helps define the nature of a machine beyond its hardware.

5 Conclusion

I have defined the modern "stored-program" computer as a machine that not only possess the hardware characteristics suggested by the "Draft Report on the EDVAC," but also was used in a certain way. Specifically the philosophy of use must take full advantage of the ability of the machine to generate its own instructions and strive for maximum generality. Finally the machine must also practically achieve significant generality. Ultimately my definition of the modern computer is qualitative rather than a strict demarcation. Therefore any claim to which machine is the first on this basis is problematic. Still, using it I have attempted to justify the common distinction made in the history of computing literature between the modern computer and their various precursors. The justification is that this definition better reflects the trends in the history of computing. In principle it might make no difference how we define things, but in practice our definitions help determine how we summarize and organize history.

Acknowledgments. Thank you to my mother and my colleague Isaac Record for proof reading versions of this paper. I also wish to thank the Special Collections of the library of the University of North Carolina at Raleigh, where I carried out research that has been integral to this paper. Finally I must thank my anonymous reviewers for their comments and corrections.

References

1. Vernon, L.: Tools for Brains. Astounding Science-Fiction, 80–90 (July 1939)
2. Goldstine, H.H.: The Computer from Pascal to von Neumann. Princeton University Press, Princeton (1993)
3. Ceruzzi, P.: A History of Modern Computing. MIT Press, Cambridge (1998)
4. Rojas, R.: The Architecture of Konrad Zuse's Early Computing Machines. In: The First Computers: History and Architectures, pp. 237–262. MIT Press, Cambridge (2000)
5. Wikipedia: von Neumann architecture, http://en.wikipedia.org/wiki/Stored_program_computer (accessed Febuary 9, 2010)
6. Rojas, R., Hashagen, U.: The First Computers: History and Architectures. MIT Press, Cambridge (2000)
7. Campbell-Kelly, M., Aspray, W.: Computer a History of the Information Machine, 2nd edn. Westview Press, Boulder (2004)
8. Bashe, C.J., Johnson, L.R., Palmer, J.H., Pugh, E.W.: IBM's Early Computers. MIT Press, Cambridge (1986)
9. Randell, B.: The Origins of Digital Computers: Selected Papers, 3rd edn. Springer, New York (1982)
10. Randell, B.: The Origins of Computer Programming. IEEE Annals of the History of Computing 16(4), 6–14 (1994)
11. Bromley, A.G.: Stored Program Concept: The Origin of the Stored Program Concept. Technical report 274, Basser Department of Computer Science. University of Sydney, Sydney (November 1985), http://www.it.usyd.edu.au/research/tr/tr274.pdf (accessed April 28, 2010)

12. Ifrah, G.: The Universal History of Computing. John Wiley & Sons, Inc., New York (2001)
13. Backus, J.: Quoted in: The IBM Selective Sequence Electronic Calculator, http://www.columbia.edu/acis/history/ssec.html (accessed Febuary 12, 2010)
14. Brooke, A. W.: Letter to Lyle R. Johnson. A. Wayne Brooke Collection, University North Carolina at Raleigh, special collections, MC#268, Box 1, Folder #2: Section 1.2 (December 27, 1980)
15. Brooke, A.W.: SSEC, the first electronic computer. Manuscript. A. Wayne Brooke Collection, University North Carolina at Raleigh, special collections, MC#268, Box 1, Folder #9: Series 2.3 writings (undated)
16. Stanford Encyclopedia of Philosophy: Existentialism: Existence Proceeds Essence, http://plato.stanford.edu/entries/existentialism/#ExiPreEss (accessed Febuary 13, 2010)
17. Campbell, S.: The Premise of Computer Science: Establishing Modern Computing at the University of Toronto (1945-1964). PhD Thesis, University of Toronto, IHPST (2006)
18. Brooke, A.W.: The Hallowed "Stored-Program Concept". Manuscript. A. Wayne Brooke Collection, University North Carolina at Raleigh, special collections, MC#268, Box 1, Folder #11: Series 2.4 writings (undated)
19. Mahoney, M.S.: The histories of computing(s). Interdisciplinary Science Reviews 30(2), 119–135 (2005)
20. Eckert, W.J., Jones, R.B., Clark, H.K.: Construction of the Lunar Ephemeris. In: Improved Lunar Ephemeris: 1952-1959, pp. 283–363. U.S. Government Printing Office, Washington (1954)
21. Grosch, H.R.J.: Computer: Bit Slices from a Life, 1st edn. Third Millennium Books/Underwood-Miller, Novato (1991)
22. Dubcombe, R.L.: Early Applications of Computer Technology to Dynamical Astronomy. Celestial Mechanics 45, 1–9 (1989)

Recession, S-Curves and Digital Equipment Corporation

David T. Goodwin and Roger G. Johnson

Birkbeck College, University of London
Dave.goodwin@gmail.com, rgj@dcs.bbk.ac.uk

Abstract. Digital Equipment Corporation (DEC) was founded in 1957 by two MIT engineers. By 1988 it had grown to be the world's second largest computer corporation. From this heady height it took a mere 10 years for the company to disappear completely. This paper looks at DEC both in relation to the S-curve of technology and how it conformed to this model in the first thirty years but missed out on the disruptive technology of PCs and workstations in the late 1980s. Also how they did not see the wave in the late 1990s and missed the opportunity to lead the market once again.

Keywords: S-curve, Digital Equipment Corporation, DEC, recession, exemplar.

Michael Mahoney wrote a number of papers[1] on the History of Computing and one thing he was always urging researchers and historical authors to do was to capture the history of failed computer companies as these are not usually written up and their archives are often destroyed, especially if they are US based. This paper goes some way to realising that goal.

Digital Equipment Corporation (DEC) was founded in 1957 by two MIT research engineers, Kenneth Olsen and Harlen Andersen. They obtained a loan of $70,000 from ARD one of the first venture capital companies, led by General Georges Doriot. Olsen created the business plan from books he had read and this plan is now housed in the Ken Olsen archives at Gordon College along with many of Olsen's memos. DEC was the jewel in the crown for ARD, making it more than $355million. In its first thirty years DEC became the second largest computer manufacturer worldwide. However over the next ten years it declined spectacularly to be taken over by a PC manufacturer. The reasons for its decline are multiple and interrelated as opposed to Schein's [1] straightforward view that it was the lack of the money gene in DEC management and its cultural DNA that brought about its downfall. Certainly, Olsen was not driven by profit, he was driven by technical excellence which defined the company direction.

DECs early history is not one without its problems, it had to survive a number of trying times as competitors rose to challenge its traditional markets. It also had to resist a number of takeover attempts from companies such as A T & T. Each time DEC emerged a stronger company except for the final time. Their growth can be linked to disruptive technology, the 'S-curve' and also to the world's financial 'wave'. The world went through four recessions and the US six, including one double-dip in the

---

[1] http://www.princeton.edu/~mike/computing.html accessed Nov 2009.

A. Tatnall (Ed.): HC 2010, IFIP AICT 325, pp. 179–188, 2010.

early 1980s, since the DEC was founded to the time that DEC was sold to Compaq. The data in table 1 is from the National Bureau of Economic Research (NBER). The NBER is considered the official arbiter of recessions, but

> *"the NBER does not define a recession in terms of two consecutive quarters of decline in real GDP. Rather, a recession is a significant decline in economic activity spread across the economy, lasting more than a few months, normally visible in real GDP, real income, employment, industrial production, and wholesale-retail sales"* [2].

Table 1. Recessions of DEC period generated from NBER data[3]

Date	Duration	Time since last recession	Peak unemployment
Aug. 1957-April 1958	8	39 months	7.5%
April 1960-Feb. 1961	10	24 months	7.1%
Dec. 1969-Nov. 1970	11	106 months	6.1%
Nov. 1973-March 1975	16	36 months	9.0%
Jan. 1980-July 1980	6	58 months	7.8%
July 1981-Nov. 1982	16	12 months	10.8%
July 1990-March 1991	8	94 months	7.8%

Source: NBER

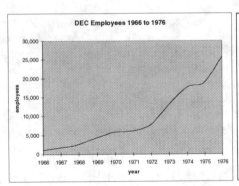

Fig. 1. DEC employees 1966-1976   **Fig. 2.** DEC net profit 1966-1976

There are a number of different recession shapes that occur in literature the most common being V-shaped, U-shaped and W-shaped. DEC suffered, like most other companies, in these recessions but shielded its workers from redundancy by redeployment

---

[2] National Bureau of Economic Research, http://www.nber.org/
[3] National Bureau of Economic Research, http://www.nber.org/cycles/cyclesmain.html

until the late 1980's when economic circumstances and pressure from Wall Street forced it to resort to large scale layoffs. In the 1970, 1973 recessions, DEC reduced its hiring, stabilised its workforce and rode out the recession as can be seen in figure 1. The result of the recessions on profit during this period is shown in figure 2 and clearly demonstrates the impact of the recessions on the profit of the company and how it impacted growth.

In the recession of the early 1980s it appears that DEC did not apply the same rules on hiring as can be seen in figure 3. This impacted the company in 1983. Ken Olsen said in a speech to the Newmans Society in 1982 that he had said publically "'*DEC didn't need recessions to straighten us out*,' but that it wasn't true, recessions made DEC strong".

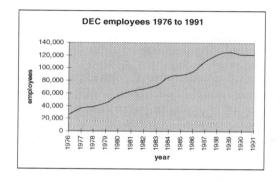

Fig. 3. DEC employees 1976-1991

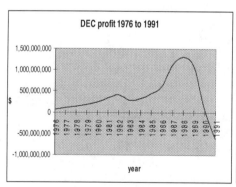

Fig. 4. DEC profit 1976-1991

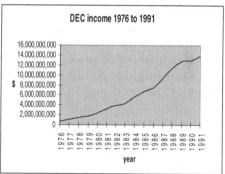

Fig. 5. DEC income 1976-1991

Again, this statement was true up until 1982 when there was no hiring freeze imposed and so expense grew and profits suffered for a number of years as can be seen in figures 4 and 5. In the mid 1980's DEC was forced into staff reductions. The method they used is described in Allen and Scott Morton [2] who did a study of employment security at DEC. It showed how a firm could manage its workforce without enforced redundancies thus maintaining its reputation for employment security. The major proportion of the reduction was in the manufacturing areas, primarily in the US. This was a forced reaction to the sudden decline in the stock market value of the company in 1983 when the value of the stock dropped by 29% in three weeks due to reporting problems within the company. They finally imposed a hiring freeze, retrained 4000 new manufacturing personnel and only had to make 600 redundant as illustrated by Rifkin and Harrar [3]. This was undertaken over a three year period from 1983.

The theory of the S-curve suggests that all businesses follow an S-curve in their development taking a certain amount of time to get to 10% market share and a similar amount of time to reach 90%. At which point, the company will fade away unless they can re-invent their product/industry and begin a new S-curve. Modus [4] states that:

"The projected life cycle of consumer products and the rate at which substitute products will gain market share is of vital interest to any company". He suggests that *"Business, in the form of products, companies and entire industries, goes through five cycles which align with the S-curve".*

He also suggests that the S-shaped curve also shows up in other life cycles. For example he states that :

- *A product S-curve may typically have a life cycle of 6 quarters.*
- *A product family S-curve, consisting of a set of related products, will typically have a life cycle of around 5 years.*
- *Basic technologies or industry S-curves, consisting of a number of product families and associated companies, typically have a life cycle of approx. 15 years.*

DEC followed this cycle successfully for many years. For product life cycles, they released a new major product almost every year from 1965, thus having overlapping S-curves. For family product lifecycles they released the PDP-8 in 1965. The last model in the family was produced in 1979 and they sold over 50,000 systems. There were 10 different models released in the 15 years. This is accepted as the first real minicomputer and heralded the start of affordable computing. Gordon Bell and Ed de Castro are credited as being the main designers of the PDP-8. Five years later in 1970 they followed it with the PDP-11, with the last product in the family being released in 1990. The PDP-11 family, excluding the 32 bit extensions, consisted of at least 23 models and was the leader in the minicomputer market for many years. In 1975 DEC released the 11/70, this was meant to be a stop-gap machine with 1000 planned. Eventually 10,000 were sold. Then, in 1977 the 11/780, DEC's first 32-bit machine, was released. In 1984 the VAX 8600 came out, in 1990 they released the VAX 9000 and finally in 1992 DEC released the 64 bit Alpha family. By the time DEC was taken over by Compaq, the VAX family consisted of around 135 models with the final VAX, in the Alpha range, being manufactured in 2005. These cycles fed DEC's incredible growth over the years.

Christensen [5] looks at technology S-curves and asks how value networks and the concept of S-curves relate to each other. He postulates that disruptive technology does not fall into the normal S-curve as it gets its commercial start in emerging value networks before invading established networks. Clearly DEC had disruptive technology with its minicomputer products, taking the mainframe makers by surprise and creating a new market for their product. With the PDP-11 they were forced into the market by Data General, which was formed by three ex-DEC engineers led by Edson de Castro who was disillusioned by DEC's decision not to go ahead with the 16 bit system he had been designing. Having formed Data General he brought a system to market very quickly and forced DEC to respond with their very successful PDP-11 range. In this instance it was Data General who had the disruptive technology that forced a reaction from DEC. De Castro had worked on the design of the PDP-8 and was working on the next system codenamed PDP-X which was to be DEC's 16 bit offering. When he left to start Data General his name was effectively wiped from DECs official histories. The Data General story is told in Tracy Kidders book "the Soul of a New Machine" [6]. In Rifkin's [3] book there is commentary on whether the team that left to form Data General were working on the Nova design whilst at DEC and Olsen is quoted as saying that DEC had a copy of their log of what they were doing for their last two years at DEC.

Looking at base technologies DEC had overlapping S-curves, starting with the PDP-8 family in 1965, the PDP-11 family in 1970 (forced by Data General's release of the Eclipse). The VAX 11/780 released in 1978 and the Alpha in 1992. This was fine when there were base technology overlaps but the Alpha was fourteen years after the VAX 11/780 and so at the limit of the S-curve creating problems for DEC in the area of uptake. This is graphically exposed in the figure 6 which shows the gap in major product release during the 1980's. The effect of this was hidden from the company by good sales of the VAX, the mid life kicker of the VAX 8600 and the sustained economic climate of the 1980s. When the recession came, it hit DEC hard especially as the VAX 9000 was two years late and released when the recession was at its worst.

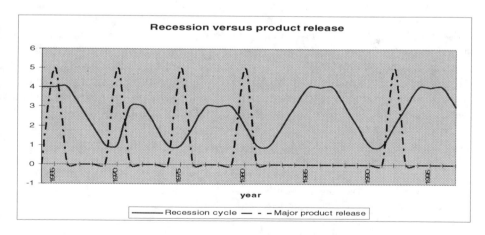

Fig. 6. Recession versus major product release

Asthana [7] looks at S-curves related to disk drive technology and comments that a phenomenon that needs careful S-curve analysis is the moving technology target. Again, this hit DEC in the early 1990s when a major development program should have put them in a leadership position in disk technology. However it took them so long to get an acceptable mean time between failure rate of their first thin-film technology disk drive that the competition had released smaller cheaper disks. This was one of the largest development projects in the company history and included a number of new technologies in the one product. DEC made a similar error in the VAX 9000 design where they again introduced three new technologies at once which caused delays in product shipment.

DEC missed two disruptive technologies in the 1980's which could have kept them on the next S-curve. The company missed the advent of the workstation by focussing on the IBM market anticipating the VAX9000 as their IBM killer. This allowed SUN amongst others to take what was once DEC's traditional market. DEC realised very late that the workstation market was important and started a workstation engineering group. This forced them into using a third party chip to challenge the competition and to cancel their own in-house project for a RISC chip codenamed PRISM. This, for a time, gave them a successful workstation and market share but it was short lived as they decided that an in-house chip was needed and started the Alpha project which confused their MIPS customers and led to a loss of marketshare. The other disruptive technology was the PC where DEC tried to create three products to compete in the marketplace when one would have given them a lead had they realised. They set up three competing groups to build a PC, the Robin, running both DOS and CP/M, a proprietary system, the Professional, and a word processing system. The groups didn't appear to know of each others existence and did not use industry standard parts so were not compatible with each other or the IBM PC standard. The sales force were confused as to which was the PC competitor and missed out on sales of the Robin PC by putting the Professional forward as DEC's main offering.

DEC is quoted in Bower and Christensen's article on 'Catching the Wave' [8] as almost completely missing the disruptive technology of the personal computer. They blame arrogance, tired executive blood, poor planning and strangely 'staying close to their customer'. Many however contend that DEC were not in a position to take on the PC market as their processes were aligned to medium volume, high margin products and not the high volume low margin market. Had they recognised the workstation market then there might not have been a crisis of confidence a few years later.

In the mid 1990s there were two other disruptive technologies that DEC had a chance to lead the market with had the company not been fighting for survival and not focussing on building for growth. The first was fast networks linked to the requirements of the internet where DEC was in a lead position in gigabit technology until Palmer sold off the network business in 1977 to concentrate on 'core' products. The second was the internet and all that that brought. DEC was the leader in internet business, forming an internet business unit under Rose Ann Giordano and creating some excellent products. DEC, according to staff interviewed, were aware of the 'wave', DEC management was always talking about riding the wave. However as figure 7 shows graphically they missed the wave of the 1980's, the Alpha S-curve was late in starting and they didn't really get back on track until the internet wave of the mid 1990's. Unfortunately, by this time the board had removed the CEO installing someone who didn't understand the internet and who was in the process of finding a merger/buyer for the company. Had he taken time to look at what was happening in 1997 things might have been different. It is clear that DEC had once again got back onto the S-curve and were well placed to ride the next wave. DEC had created Alta Vista in its Western Research Lab as a way to use the power of the Alpha processor and had already established the Alpha as a force in internet business suppliers such as Amazon because of its power. This was mainly on Unix based systems however rather than Open VMS.

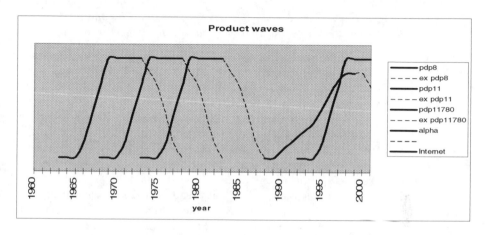

Fig. 7. New product 'waves'

Russ Jones in a chapter of Cronin's book [9] looked at DEC's internet business and its leadership position in late 1994. DEC was the first fortune 500 company to have its own web site when it opened the first commercial home page on the internet in October 1993. They had the majority of the business server market in the internet arena with Amazon as a major customer. When they released AltaVista it was an instant hit and the name went from nothing to worldwide fame in six months being better known than DEC itself. It was the search engine used by Yahoo, cementing its position as search engine of choice. DEC produced the first internet firewall product, the first tunnelling software in 1991 and was well ahead of the competition.

When the founders of Google, Larry Page and Sergey Brin, came to DEC with an offer of joining with AltaVista for $1 million, DEC's response was negative due to a 'not invented here' attitude and senior management preparing for the sale of the company. This was certainly another opportunity missed for DEC. Palmer didn't understand what he had in AltaVista. He didn't understand the potential of the internet, valuing AltaVista at $0 when the sale to Compaq went through. In 1999 Compaq sold AltaVista to CMGI for $2.3 billion. Bell in his appendix to Schein [1] stated that Internet business products were perfect for DEC, they had all the pieces including servers, software and networking, however they didn't understand how to organise to engage in a new market.

DEC, IBM and HP are all exemplars quoted by Peters and Waterman [10] as organisations with structures and strategies that are the ones to follow. Yet by 1990 all three were in deep trouble suggesting that the study by Peters and Waterman [10] was in some way inaccurate. In their updated version they add an authors note on Excellence 2003 where they try to justify their publication in terms of excellence value. They totally ignore DEC in this commentary. However a paper written by Crainer and Dearlove [11] analyses the companies in the book ten years later and report that Michelle Clayman found that

'the companies featured in the book beat the stock market by one percent, whereas the mass of unexcellent companies beat the stock market by around 12% over the five years following the book's publication'.

Sheth [12](page 4) comments that DEC was a fun place to work and suggest it was no wonder that Peters and Waterman [10] considered DEC as on of the 15 exemplars. However, he goes on to question DECs status at the end of the decade and its late entry into the PC and workstation market, going on the describe the company as one where executives were fleeing, and layoffs, once an abhorrent practice in DEC, were now occurring.

McGrath [13] wrote of the product strategies of high technology companies and has a number of extremely pertinent comments to the situation the DEC found itself in the 90's. He explores the importance of strategy, and the need for changes in strategy as technology changes, the reaction to stagnation of strategy and the potential for diversification. He comments on the selection of Palmer as the CEO and also looks at some of the products that DEC had and their potential for strategic advantage that was not followed up on. He also has commentary on many of the competitive companies at the time and their strategies which either helped them survive or aided their demise. McGrath [13] implies that DEC strategy on the Internet was only developed by Palmer in 1997 which was, in his opinion a few years too late. This was taken from the company report of 1997. However this strategy had been developed a few years earlier within the company. Rose Ann Giordano had been made VP of the Internet Business Group in 1994 to develop the vision and strategy but this was not recognised as strategic by Palmer until later, many saying that he did not understand the value of the Alta Vista product range.

Pettigrew, Thomas and Whittington [14] talk of the diversification index for the Fortune 500 companies declining from 1.0 to 0.67 in the period 1980 to 1990 as divestiture replaced diversification driven by the shifting of corporate goals from growth to profitability and pressure from shareholders and financial markets. This led to the ousting of many CEO's, including Ken Olsen, by increasingly independent boards. The move from diversification appears to go against some of the findings as quoted by Pettigrew, Thomas and Whittington [14] where related diversification linked closely to core business was superior to unrelated diversification.

General Georges Doriot in the address to the Newcomen Society[4] stated that when you have a strong president your directors should be very peaceful. In fact DEC directors were placid for most of DECs history. Often board meetings were said to be more of a social event than a formal board[5]. The board members were weak according to

---

[4] Digital Equipment Corporation, the First Twenty-five Years, Kenneth Olsen speech to the Newcomen Society, Newcomen Publication Number 1179, Sept 21st 1982.
[5] Interview with former board member in Boston, October 2009.

many commentators and most directors, when contacted, were unwilling to talk about their time on the board of DEC. Many did not understand the technology or the business but still made decisions that impacted the company direction. One member that did agree to discuss their time on the board commented that the decision to replace Olsen was not voted on. He also stated that they decided on Palmer without looking outside the company as he appeared to understand the PC business. Their decision was based on videos they had asked senior managers to complete stating what they would do to rescue the company. Palmer took coaching in video techniques prior to recording his video.[6] Board meetings became increasingly acrimonious once DEC's profits declined and General Doriot died, leaving Olsen isolated. Olsen tried in vain for many years to get his senior managers to give him realistic budgets to give to the board.[7] In 1992 the board asked for an austerity budget and Olsen asked for a budget for growth. This clash plus a reluctance to cut as many heads as the board had asked for led to the board asking Olsen to leave.

In conclusion, one of the many factors contributing to DEC's downfall was missing the S-curve in the mid 1980s and not having a strategy to recover. They left it too late to move to RISC architecture allowing competitors to capture the workstation market. They were riding the wave during the 1960's and 1970's but their success made them try to get into larger markets rather than their traditional ones. They missed the PC and workstation revolution by focusing on IBM's business and their profits suffered as margins were eroded. However, they could have recovered in the late 1990s had they realised that they were on the next S-curve and a leader in the field of the internet. The Board's decision to select Palmer to succeed Olsen was taken in haste and with little apparent though as to what the company direction should be. He was chosen mainly because the board though Olsen had missed the PC revolution.

References

1. Schein Edgar, H., DeLisi, P.S., Kampas Paul, J., Sonduck Michael, M.: DEC Is Dead, Long Live DEC: The Lasting Legacy of Digital Equipment Corporation. Berrett-Koehler, San Francisco (2003)
2. Allen Thomas, J., Scott Morton Michael, S.: Information Technology and the Corporation of the 1990s. Oxford University Press, Oxford (1994)
3. Glenn, R., George, H.: The Ultimate Entrepreneur: The Story of Ken Olsen and Digital Equipment Corporation. Contemporary Books, Chicago (1988)
4. Theodore, M.: Conquering Uncertainty: Understanding Corporate Cycles & Positioning Your Company to Survive the Changing Environment. McGraw-Hill, New York (1998)
5. Christensen Clayton, M.: The Innovator's Dilemma. Harper Collins, New York (2006)
6. Tracy, K.: The Soul of a New Machine. Back Bay Books, Boston (2000)
7. Praven, A.: Jumping the Technology S-Curve. IEEE Spectrum, 49–54 (June 1995)
8. Bower Joseph L., Christensen Clayton M.: Disruptive Technologies; Riding the Wave. Harvard Business Review, 43–53 (January-February 1995)

---

[6] Interview with Olsen's personal assistant October 2009.
[7] Ken Olsen's memos in the Ken Olsen Archives at Gordon College, Massachusetts.

 9. Cronin Mary, J.: The Internet Strategy Handbook; Lessons from the New Frontier of Business. Harvard Business School Press, Boston (1996)
10. Tom, P., Waterman Robert Jr., H.: Search of Excellence: Lessons from America's Best Run Companies. Profile Books, London (2004)
11. Crainer, S., Dearlove, D.: Excellence Revisited. Business Strategy Review 13(1), 13–19 (2002)
12. Sheth Jagdish, N.: The Self Destructive Habits of Good Companies: and How to Break Them. Wharton School Publishing, New Jersey (2007)
13. McGrath Michael, E.: Product Strategy for High-Technology Companies: Accelerating your Business to Web Speed. McGraw Hill, New York (2000)
14. Andrew, P., Howard, T., Richard, W.: Handbook of Strategy and Management. Sage Publications, London (2007)

ETHICS: The Past, Present and Future of Socio-Technical Systems Design

Shona Leitch and Matthew J. Warren

School of Information Systems,Faculty of Business and Law,
Deakin University, Burwood, Victoria, Australia, 3125
matthew.warren@deakin.edu.au

Abstract. Since computers were first introduced in the late 1960's there has been continued debate on the impact of technology, organisations and staff within those organisations. Enid Mumford was one of the key researchers who looked at the Socio-Technical implications through the decades, and as part of her research she developed the ETHICS method to help improve the integration of technology in organisations and society.

Keywords: Enid Mumford, ETHICS, participation, and systems design.

1 Introduction

Across the globe, there are now many different types of Information Systems in place, from databases, expert systems, cloud computing application, transaction processing systems, Internet based systems, decision support systems, etc. The use of these systems is very different from global Internet based systems, organisational systems to personal systems where the number of users can vary from three hundred million users to a single user.

The computer revolution was predicted in the late 60's and during the next thirty years we saw the introduction of corporate computer systems, personal computer systems, home micro computers and the development of the Internet. Enid Mumford understood the impact that technology and systems would have upon us all and her research over the decades focused upon the issue of this impact; on organisations as well as on the individual.

This paper will explore the early social-technical research, the development of the ETHICS model and how this has changed over the years and includes a discussion of how Enid Mumford's research could influence future research areas and focus.

2 History of Socio Technical Design

According to Mumford and Beekman (1994), the Socio-Technical system design was the product of a group of social scientists who came together at the end of the Second World War and formed the Tavistock Institute of Human Relations in London. The Tavistock Institute was established in 1946 by this group, many of whom had collaborated in wartime projects and most of whom had been members of the Tavistock Clinic before the war. The Tavistock Clinic was a therapeutic establishment

A. Tatnall (Ed.): HC 2010, IFIP AICT 325, pp. 189–197, 2010.

concerned with mental health and individual development and this was also the initial focus of the members of the Institute, although they were applying their ideas to workers in industry. In 1949, the Tavistock Institute made its first major contribution to the theory of Socio-Technical design with a number of field projects in the British coal industry.

Mumford and Beekman (1994) identified that the major outcome of the early Socio-Technical research was:

"If a technical system is created at the expense of a social system, the results obtained will be sub-optimal."

3 Development of ETHICS

Based upon Mumford's Socio-Technical experiences she developed the the participational method known as ETHICS (Effective Technical and Human Implementation of Computer based System). The work on ETHICS was undertaken by Professor Enid Mumford during her time at Manchester Business School, UK (Mumford, 1983a).

ETHICS is a participational (also referred to as a Socio-Technical approach) approach that focuses upon people and procedures. This Socio-Technical approach is defined by Mumford as "one which recognises the interaction of technology and people and produces work systems which are both technically efficient and have social characteristics which lead to high job satisfaction" (Mumford, 1983b).

The ETHICS approach was based on the observation of failure of many systems which followed more traditional route of considering technical and economic factors (Davis et al, 1992). One of the key motivating questions was whether or not analysts and designers held a view of users that was different from that held by the users of information systems. From these observations, Mumford concluded that the development of information systems is not a purely technical issue, but an organisational issue which is fundamentally concerned with the process of change (Mumford and Weir, 1979; Mumford, 1995).

The use of participation allows users to have some level of contribution in the system development life cycle, this participation often take the form of single representatives. The user participant is often called upon after the major decisions have been taken; this limits the user participation of involvement within the system development (Nurminen, 1988).

The original ETHICS methods were developed in the UK in the late 1960's to deal with the impending information revolution (Mumford and Ward, 1968) of the 1970's. The early conceptual models of ETHICS were concerned with: ensuring users were satisfied with their jobs and trying to determine the impact that computers could have upon their job; and the perception that computers were perceived as agents of change within organisations.

These principles were used as the foundation of the formalised ETHICS method. Around this time Mumford (1969) examined the impact of implementing computers within organisations, and determined that the successful introduction of technical changes required: the use of interdisciplinary planning teams, particularly when goals and objectives are being defined; awareness of the fact that technical changes have secondary as well as primary consequences; and planning does not take place in a static situation.

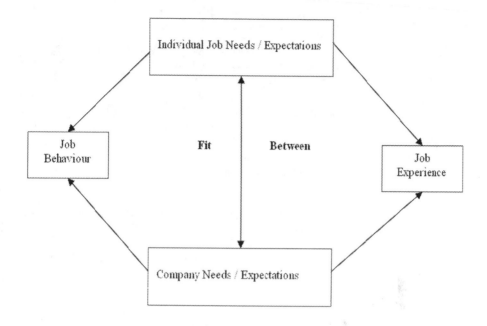

Fig. 1. Conceptual model describing the benefits of technology

The original systems that were being evaluated using this approach were office computer systems and the impact that their introduction would have on office clerks (Mumford and Banks, 1967). Much of this earlier research was based upon trying to determine the impact of these newer technologies, such as micro-computers upon organisations. By the late 1970's the use of technology within organisations was more common and were becoming formalised and we started to see models being developed (Legge and Mumford, 1978) to describe complex issues. This research went beyond the simple analysis of impact on an organisation and began to considering the effects that the use of computers would have on the individuals within the organisation, including changes to job roles, individuals' perceptions, behaviour and needs, expectations and job satisfaction.   An example of one of Mumford's earlier models relating to the benefits of technology is shown by Figure 1 (above).

A key area in her earlier research is the concept of job satisfaction; Mumford and Weir (1979) define job satisfaction as:

the attainment of a good "fit" between what the employee is seeking from his work - his job needs, expectations and aspirations - and what he is required to do in his job - the organisational job requirements which mould his experience.

This definition draws upon earlier work looking at the job satisfaction of computer specialists (Mumford, 1972). A continuation of the research by Mumford saw continued development in the key area of participation and how different forms of participation could be used within the ETHICS method. Mumford (Mumford and Henshall, 1979) defined the following levels of participation:

1. *Consultative* – This is when an existing body, e.g. steering committee, is used to implement the change process. This committee would then consult users on the effect that change would have upon them;
2. *Representative* – This is when a cross selection of users affected by change, are brought together into a design group. This ensures that representatives effected by change have the same powers in the committee as those bringing about change; and;
3. *Consensus* – This is when all the staff impacted by the change are involved in the design process. Representatives of the staff are elected to form the design committee.

Another key area is the unique view of the Socio-Technical approach. Mumford (1983a) redefines the Socio-Technical approach as:

"one which recognises the interaction of technology and people and produces work systems which are both technically efficient and have social characteristics which lead to high job satisfaction."

The research undertaken by Mumford is encapsulated in the ETHICS methodology (Mumford and Weir, 1979) to implement system design. The earlier ETHICS methodology consisted of seven stages, which are (Mumford and Weir, 1979):

Step1 - *Diagnosis:* Determine the information required for the diagnosis of human needs, collected through the use of questionnaires. The results of the survey are analysed to determine user needs, the new system should be designed to meet user requirements, as far as possible;

Step2 – *Socio-Technical system design*: Define the human objectives, which the new system should achieve, based on the social diagnosis of step 1;

Step 3 – *Setting out alternative solutions*: Define the possible social and technical solutions in order to achieve the desired requirements of step 1 and step 3;

Step 4 – *Setting out possible Socio-Technical solutions*: Combine the separate social and technical solution into a combined list of solutions;

Step 5 – *Ranking Socio-Technical solutions*: List the Social-Technical solutions which achieve the objectives set in step 2 and cater for the human needs as defined within step 1;

Step 6 – *Preparing a detailed work design*: Develop system specifications and work plans for the top choices from step 5;

Step 7 – *Accept the best possible Social Technical solutions*: Evaluate the plans from step 6 and implement the best possible Socio-Technical solution.

Committees of individual users, managers and IT staff would be the ones who would conduct the different stages of the ETHICS methodology. The original ETHICS methodology was extended to take into consideration such issues as availability and reliability of the systems once they have been introduced. The introduction of new technology into an organisation can also be thought of as a human issue, relating to (Mumford, 1995):

User requirements: New technology directly affects users. There is little evidence that managers have recognised the need of using IT to change the way they do business. User requirements should be incorporated fully into the system design from the start so that the system that is designed actually complies with user requirements; and

User job satisfaction: The way in which a computer can have a direct effect upon the user and the way they use the system. If the user is unsatisfied with the system they will become less motivated and users will take longer to carry out tasks, or might not even use the system at all.

It was during the 1980's that micro computers began to have an obvious impact upon organisations. Mumford undertook a number of projects in the 1980's using ETHICS to redesign administrational support systems, in particular secretary systems (Mumford, 1983c). Also during this time ETHICS was used to develop unusual systems such as an expert system for Digitial Equipment Corporation (DEC), the XSEL system was developed for their sales office to help configure DEC hardware system for customers (Mumford and MacDonald, 1989). The ETHICS principles were also used to determine the value system of large organisations (Mumford, 1981).

During the eighties, the ETHICS methodology was expanded to fifteen levels (Mumford, 1986), the stages were:

Stage 1 – Why Change? Determine whether there is need for change;

Stage 2 – System Boundaries: Identify the boundaries of the system that has to be developed;

Stage 3 – Description of existing systems: Determine how the existing system works looking at issues such as the sequence of events within that system;

Stage 4,5,6 – Definition of key objectives and tasks: From the analysis of the system determine what the key tasks and objectives are and related information;

Stage 7 – Diagnosis of efficiency needs: Determine possible weak links in the existing system;

Stage 8 – Diagnosis of job satisfaction needs: Determine users' perception of the current system in regards to job satisfaction. This would be carried out via the use of questionnaires. The results of the questionnaire would be drawn into the actual system design;

Stage 9 – Future Analysis: An analysis of the future requirements of the system is undertaken, this is to ensure that the system design covers possible areas of potential change;

Stage 10 – Specifying and weighting job satisfaction: Rank the key objectives based upon the analysis of stages 7, 8 and 9;

Stage 11 – Organisational design of the new system: Develop a design of the system that focuses upon the issues identified relating to efficiency, job satisfaction, etc (this runs in parallel with Stage 12);

Stage 12 – Technical Options: Determine the technical aspect of the system including issues such as hardware, software, human-computer interface, etc;

Stage 13 – Preparation of a detailed work design: Prepare the system plan in more detail e.g. defining data flows, responsibilities, etc;

Stage 14 – Implementation: Oversee the implementation of the work design plan;

Stage 15 – Evaluation: Evaluate the new system to ensure that it complies with the required objectives.

A number of criticisms of the ETHICS method have been expressed (Avison and Fitzgerald, 2006):

- unskilled users cannot design;
- management will not accept it; and
- it removes the right to manage from managers
- slow and costly in staff time and effort

To overcome some of these concerns over the applicability of ETHICS, a newer version of ETHICS was developed called QUICKethics (QUality Information from Considered Knowledge) (Mumford, 1993). It was developed to create and maintain management interest (Avison and Fitzgerald, 1995) and it is broken down into five main stages:

- Describe the work mission, key tasks, critical success factors and most serious problems.
- Describe the objectives, critical success factors, major problem, day-today activities, and potentials for future developments associated with each of the key tasks.
- Describe the information needs associated with these tasks in order to achieve the objectives, attain critical success factors and avoid major problems, as well as monitoring performance and understanding future developments.
- Prioritise these information needs according to which are essential and which merely desirable, and which are quantitative and which are qualitative.
- Work with others to establish an information model so that information flows through the organisation to those who require it.

(Mumford, 1983b)

Whilst the standard ETHICS Methodology was a top down, user driven approach, Mumford rejected this approach in her QUICKethics approach and concentrated the methodology as starting from the centre and forming small working groups as a core part of the process (Mumford, 1983b). By working in this way many of the benefits of participative development could be achieved but in a much shorter timeframe than under the ETHICS methodology.

Mumford also proposed using QUICKethics as part of the PROGRESS method to help in Business Process Re-Engineering. The aim of this approach was to rethink and restructure business processes so to make them more efficient, more effective in achieving business goals and more able to provide a high quality work environment that motivates employees (Mumford and Beekman, 1994).

4   Analysis of ETHICS

A common reaction to ETHICS is for researchers to say that it is impractical (Avison and Fitzgerald, 2006) due to the structured nature of the ETHICS and QUICKethics approaches. Another common criticism of the ETHICS method is that it is unworkable (Avison and Fitzgerald, 1995) as the use of committees to make decisions means that unskilled workers could make decisions about very technical applications.

Another argument against ETHICS is that it removes the rights of managers to manage, which could have dramatic impacts in the development of the system as well as cause conflict issues within the organisation. The strong focus of the ETHIC method on participation affords it many benefits as discussed earlier in the paper However this strong use of participation can also cause problems in a systems development process including:

- management/workforce distrust;
- working with managers tends to inhibit workforce;
- conflicts of interest between stakeholders/stress;
- users can't visualise rapidly developing computing possibilities;
- team working skills are required;
- consensual solutions unlikely to be radical;
- technical experts feel demoted to advisers.

The simple answer to this is that each systems development project needs to be assessed as being suitable for the use of ETHICS as in the case of any chosen methodology. One methodology may suit a particular development well whilst another would prove to be incredibly ineffective. In the "correct" situation ETHICS can provide an invigorating and dynamic experience for an entire organisation.

Mumford (2003) argues that ETHICS places emphasis on identifying new approaches to tasks and problems and new relationships within and outside the organisation and that this is the strength of ETHICS. Mumford (1996) also argues that importance of designing for the future, hence the structural approach to ensure successful design. Mumford was also interested in change and the way it is reflected in society and organisations. She accepted change as a principle that pervades modern societies and their organisations. At the same time, she believed that change is not something that must be suffered passively but that should be embraced. Change must "always be accepted by the participants" (Mumford and Ward, 1968; Stahl, 2007).

Examining the basis for Mumford's work is a refreshing change from reviewing the literature concerning many other methodologies. Where other methodologies make assumptions about the intent of the various stakeholders and concentrate on structure and process issues, Mumford is interested in values and their relationship to technology and work.

Much of Mumford's research has focussed upon the discourse between technology, organisations and staff within those organisations. These issues are as important now as they were in the late 1960's. The authors intend to carry on Enid Mumford's research into Socio-Technical approaches looking at issue in relation to Information Security and the impact that Internet based systems can have, the authors have already used a variation of ETHICS called SIM-ETHICS to assist in the implementation of security technologies within an organisation (Warren and Batten, 2002).

5 Conclusion

Enid Mumford passed away in 2006. Her research achievements were recognised by a number of international prizes. In 1983, she won the US Warnier prize for her contributions to information science research. In 1999, she won a Leo lifetime achievement award of the Association for Information Systems (The Guardian, 2006).

Enid Mumford made a major contribution to research in a number of areas, in the authors opinion one of her key research findings was Mumford and Beekman (1994):

> *"If a technical system is created at the expense of a social system, the results obtained will be sub-optimal."*

The other major contribution is that the findings that she made at the start of her research in the 1970's are still current in the early part of the twenty first century. In 1974, she identified (Hedberg and Mumford, 1974):

> *Perhaps the strongest influence that changes the practice model held by the systems designers will be when the users get up and shout "we are not as you think we are".*

This statement is as relevant now as it was then, and demonstrates that Enid Mumford's research will be applicable for many more decades to come.

References

1. Avison, D.E., Fitzgerald, G.: Information Systems Development: Methodologies, Techniques and Tool, 4th edn. McGraw-Hill Education, UK (2006)
2. Avison, D.E., Fitzgerald, G.: Information Systems Development: Methodologies, Techniques and Tools, 2nd edn. McGraw-Hill, UK (1995)
3. Davis, G.B., Lee, A.S., Nickles, K.R., Chatterjee, S., Hartung, R., Wu, Y.: Diagnosis of an information systems failure: A framework and interpretative process. Information and Management 23, 293–318 (1992)
4. Hedberg, B., Mumford, E.: The Design of Computer Systems. In: Proceedings of the IFIP Conference on Human Choice and Computers, Vienna, Austria (1974)
5. Legge, K., Mumford, E.: Designing Organisations for satisfaction and Efficiency. Gower Press, UK (1978), ISBN 0-566-02102-1
6. Mumford, E.: Computers, Planning and Personnel Management. Institute of Personnel Management, UK (1969)
7. Mumford, E.: Job Satisfaction: A study of Computer Specialist's. Longman Group Publishers, UK (1972), ISBN 0-582-45008-X
8. Mumford, E.: Values, Technology and Work. Martinus Nijhoff Publishers, The Netherlands (1981), ISBN 90-247-2562-3
9. Mumford, E.: Designing Participatively. Manchester Business School, UK (1983a), ISBN 0-903808-29-3
10. Mumford, E.: Designing Human Systems. Manchester Business School, Manchester (1983b), ISBN 0-903808-285
11. Mumford, E.: Designing Secretaries. Manchester Business School, Manchester (1983c), ISBN 0-903808-250
12. Mumford, E.: Using computers for Business. Manchester Business School, Manchester (1986)
13. Mumford, E.: Designing Human Systems For Health Care, The ETHICS Method, 4C Corporation, Netherlands (1993), ISBN 90-74687-01-6

14. Mumford, E.: Effective Requirement Analysis and Systems Design. The Ethics Method, Macmillan (1995)
15. Mumford, E.: Systems Design: Ethical Tools for Ethical Change Method. Macmillan, UK (1996)
16. Mumford, E.: Redesigning Human Systems. Information Science Publishing, Idea Group Inc., USA (2003), ISBN 1-59150-118-6
17. Mumford, E., Beekman, G.: Tools for change & Progress. CSG Publications, The Netherlands (1994), ISBN 90-75198-01-9
18. Mumford, E., Banks, O.: The Computer and the clerk. Routledge & Kegan Paul Limited, UK (1967)
19. Mumford, E., Henshall, D.: A participative approach to computer systems design. Associated Business Press, UK (1979), ISBN 0-85227-221-9
20. Mumford, E., MacDonald, W.: XSEL'S Progress: The Continuing Journey of an Expert System. John Wiley & Sons Ltd., UK (1989), ISBN 0-471-92322-2
21. Mumford, E., Ward, T.B.: Computers: Planning for People. Batsford Limited, UK (1968)
22. Mumford, E., Weir, M.: Computer systems in work design – the ETHICS method. Associated Business Press, UK (1979), ISBN 0-85227-230-8
23. Nurminen, N.: People of Computers: Three ways of Looking at Information Systems. Chartwell-Bratt, Sweden (1988), ISBN 0-86238-184-3
24. The Guardian. Obituary: Enid Mumford (2006),
    `http://www.guardian.co.uk/news/2006/may/03/guardianobituarie
    s.obituaries` (accessed January 10, 2010)
25. Stahl, B.C.: ETHICS, Morality and Critique: An Essay on Enid Mumford's Socio-Technical Approach. Journal of the Association of Information Systems 8(3) (2007), ISSN: 1536-9323
26. Warren, M., Batten, L.: Security Management: An Information Systems Setting. In: Batten, L.M., Seberry, J. (eds.) ACISP 2002. LNCS, vol. 2384, p. 257. Springer, Heidelberg (2002)

Lessons from Discarded Computer Architectures

Andrew E. Fluck

University of Tasmania
Locked Bag 1307, Launceston, Tasmania, 7250, Australia
Andrew.Fluck@utas.edu.au

Abstract. The BBC microcomputer was one of several nationally produced which were superseded by the International Business Machines (IBM) Personal Computer (PC). This reflected the results of both an international market competition and rivalry between different US processor manufacturers. Along with the hardware, valuable software and supporting educational ideologies were discarded. As we make choices about technological innovation, to what degree are we selecting potential efficacy or responding to marketing hype?

Keywords: BBC microcomputer, IBM Personal Computer, Apple II, computer hardware, operating systems.

1 Introduction

The BBC microcomputer was an 8-bit machine based on the Motorola 6502 processor. It made a huge impact in British schools, putting predecessors into the shade of its colour graphics. Its successor, the Archimedes was almost as successful, but ran into the juggernaut of the IBM PC – and the rest is history. Almost.

2 The BBC Microcomputer - A Withered Branch

In 1979-80 the British Broadcasting Company (BBC) started the BBC Computer Literacy Project and put out a tender for a microcomputer to accompany the television series The Computer Programme. Acorn was a firm started by two former Sinclair employees, marketing director Chris Curry and researcher Hermann Hauser. Their firm won the tender in April 1981 and released the BBC Microcomputer later that year. The large keyboard unit connected to a conventional television, which became the screen for the computer. Based on the 8-bit Motorola 6502 processor, the initial model had 16k bytes of RAM, and cost GBP 299. Backup storage was initially to cassette tape, with floppy disks (5¼") coming later. Notably the computer had many interfaces, including networking (CDMA econet), a serial RS-423 port, analog input (for joysticks etc.), parallel input/output user port Centronix printer port, RGB, composite video and TV outputs. The operating system was in read-only memory (ROM), and this contained a BASIC interpreter. Additional language ROMS could be installed to give extra functionality. One such language was Micro PROLOG which was

A. Tatnall (Ed.): HC 2010, IFIP AICT 325, pp. 198–205, 2010.
© IFIP International Federation for Information Processing 2010

released by Acornsoft in 1985. This was a declarative language, very different from the sequential algorithmic languages familiar to most programmers of the time. To overcome this novelty, a Man in the Street Interface (MITSI) was written by Jonathan Briggs at Imperial College London, and made popular by Jon Nichol from the School of Education, University of Exeter [1,2]. Jackie Dean worked in Western Australia, providing an antipodean link.

The BBC micro was put on sale, and additionally the British government Department of Trade and Industry arranged to place one into every school, since the advent of microelectronics was expected to have a major impact on commerce and work. The Department of Education set up a series of national advisory units which continue today as Becta (formerly the British Educational Communications and Technology Agency). Production was discontinued in 1994 by which date over one million BBC Micros had been sold in the UK and Europe [3].

Fig. 1. Logo of the BBC Computer Literacy Project

John Coll, an electronics teacher from Oundle School was hired to write the user manual for the BBC micro [4] and also appeared in related television programs. This author recalls sitting in his office to discuss a new programming project when one of the operating system programmers rushed in. He had recoded some graphics routines and saved 10 bytes of space in the ROM. This would make possible the inclusion of an additional function!

The BBC micro was very popular in British schools. As part of the government support for schools IBM sponsored a project for the Redbridge SEMERC (Special Education Microelectronics Resource Centre) by M-Tec computer services (UK) to create a card for the IBM Personal Computer which would replicate some of the interface ports on the BBC micro. Devices such as the concept keyboard and various robotic turtles (controlled by variants of the LOGO language) were so popular, this sought to ease the conversion of educational software onto the more dominant platform. The SNIC card (special needs interface card) had a short life.

Acorn went on to produce successors to the BBC micro: the Archimedes and then the RISC-PC in 1994. However, the company was broken up in 1998, and Castle Technology acquired the rights to market and produce this later machine [5]. The RISC-PC was supplanted by the Iyonix PC in 2003, but even this was discontinued in 2008 [6].

One thing that did emerge from these discarded Acorn computer architectures was the idea of a RISC (reduced instruction set computer). The first such processor was incorporated in Acorn's 32-bit computer in 1985 – the Archimedes [7]. A new company, Advanced RISC Machines (ARM) was formed in 1990, and became very successful in designing this new kind of chip. ARM processors had very low power consumption, and were ideal for mobile application where battery life matters. Leading into 2007 they were used in over 98% of the world's billion mobile phones [8].

A person who links the withered branch of the BBC computer to the highly successful ARM processor tree is Tudor Brown. He studied Electrical Engineering at Cambridge University, and was awarded an M.A. in Electrical Sciences. He was enticed back to the city again in 1983 to join Acorn Computers, where he worked on the ARM R&D program as Principal Engineer. When ARM spun out from Acorn as a joint venture with Apple, he became Engineering Director and then Chief Technical Officer from 1993 [27] In October 2000 he was appointed Executive Vice President, Global Development and in October 2001, joined the board of ARM as Chief Operating Officer. He became President in 2008 with responsibility for developing high-level relationships with industry partners and governmental agencies and for regional development [9].

This withered branch of computer architecture contains the seeds of a comparison between marketing success and intellectual striving. It embraces the success of algorithmic thinking based upon uni-processor machines and a failure to make progress with multi-processor computers. To explain the next withered branch, we need to give an overview of the instruction sets associated with particular processor families.

3   Links between Processor Families and Operating Systems

An operating system is a collection of programming codes designed to provide a consistent interface between hardware and the software applications run by the computer user. In this sense, an operating system can run on any hardware to which it has been adapted, and the software application will run as expected. When an operating system runs consistently on more than one processor, it must be coded using a different instruction set for each. The instruction set consists of all the various instructions that the processor can execute.

It can be quite difficult to create and maintain an operating system, and this causes some inertia in the versioning process to cope with different processors. Therefore while it is not strictly necessary for a particular operating system to be associated with a specific processor development family, in the main this has been the case.

The two most significant such associations have been Windows with Intel x86 line processors and Apple's Macintosh operating system with Motorola 6xxx line of processors and others. The Microsoft Disk Operating System (MS-DOS) with its successive versions of Microsoft Windows has been developed to run on 16, 32bit and 64bit processors from Intel; currently the Atom and Core-i7. These have been developed from the original 8086 (16bits, 1978), through the 80486 and Pentium (in 1993) versions [10].

The Apple line of succession began in 1977 with the Apple II using the Motorola 6502 processor which ran operating systems such as CP/M [11]. The Apple II was the first true "personal computer" which was factory built, inexpensive and easy to learn and use. Provided with the most extensive set of software and low cost floppy disks, the Apple II was also the first personal computer capable of color graphics and easy modem operation. Development of the Visicalc spreadsheet program created a business tool that made adoption of Apple II a regular part of business [12]. The processor development series moved to the Motorola 68000 and the Apple Macintosh line of computers. These ran a new operating system with some routines in Read Only Memory (ROM) for speed. This operating system was called 'System' in 1984, but gradually became called MacOS. It incorporated elements from FreeBSD's and NetBSD's implementation of Unix from 1996 [13].

The processors used by Apple changed from the Motorola 68040 series to PowerPC chips (Motorola and IBM) and then to Intel x86 chips from 2007.

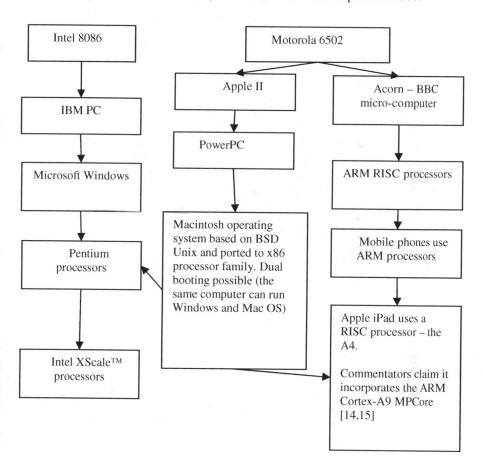

Fig. 2. Development pathways for two dominant personal computer families

It should be noted that all these processor families were von Neumann computers – with a single processor. This architecture has a single memory store which holds instruction codes and data. Programs run sequentially, and therefore an algorithmic approach is highly congruent with such machines. Fortunately many problems can be solved by such methods, and the high speed (3GHz is not uncommon) of processor operation supports a wide range of useful functions.

An alternative was offered by the *Transputer* from Inmos [16] from 1983 [17]. Inmos Limited was a British semiconductor company, founded by Iann Barron, based in

Bristol and incorporated in November 1978. Inmos ceased trading, and many staff moved to SGS-Thomson (now STMicroelectronics) in April 1989 and was fully absorbed by 1994 [18].

Fig. 3. A Transputer chip [19]

4   Alternatives to the Von Neuman Design – Triumph of Algorithm over Heuristic

Each Inmos Transputer chip contained a processor, RAM storage and communication ports. The chips were designed to be wired together in arrays. Whereas most single processor computers are limited to handling problems using sequential steps, parallel hardware could solve problems faster by undertaking multiple computations simultaneously.

The question is, how can programmed solutions be stated in this new form? If a sequential language is used (such as the old FORTRAN code), a translator must identify and exploit steps that can run simultaneously [20]. If a non-sequential language is used instead, a different kind of approach is needed.

Some problems, such as weather forecasting or industrial product design, use mathematical methods such as finite element mesh analysis [21], which map very well onto processor arrays. Of course, using very high speed networks, the computer array can be established as a virtual machine, and therefore multiple PCs can be used for similar purposes using control software such as Beowulf clusters [22]. Explorations into distributed processing architectures still continue [23].

The general application of parallel processor architectures depends upon theoretical advances that can establish the importance of parallel universality [24].

Others have investigated parallel processor arrays. These include Intel, whose iWARP product was evaluated for use in the international space station [25] where it gave computational speed increases of an order of magnitude over a single processor equivalent. Product development led to a commercial supercomputer product, the Intel Paragon XP/S [26].

Despite these advances, parallel computing has not achieved the hopes of the fifth generation computing project, and applications appear to remain restricted to specialised (but important) situations. Interest has moved to quantum computing, where the superimposition of energy states are expected to be processed using optical methods to solve problems extremely quickly.

5 Conclusion

There appear to be four lessons to be learned from this story of discarded computer architectures. Firstly, the BBC microcomputer was quickly supplanted by the open-architecture IBM PC and the closed architecture Apple Macintosh. However, the RISC processor designed for the BBC microcomputer's successor has been widely adopted because of its very low power consumption. In a world faced by climate challenge and a huge growth in the use of mobile computing devices, this has been a winning strategy.

The second lesson can be drawn from the way personal computer operating systems have been largely linked to processor families. Most of the main lines of growth have been limited to von Neuman uni-processor architectures. Even the development of quad or more core processors are still just variants of this sequential flow machine, albeit allowing a few more threads to access different parts of memory simultaneously. The lesson of how to break away from this design template has yet to be learned.

The third lesson relates to the Transputer and ways in which these chips could be wired in arrays. Even declarative languages such as Prolog were not ported successfully to this architecture in such a way to facilitate problem solving which became mainstream. We appear to lack an understanding of how to implement solutions using these kinds of techniques.

Finally, our hopes of leapfrogging this difficulty rest with novel computing techniques such as quantum or biological constructs. It remains to be seen how successful these will be – and they may need to take heed of the other three lessons to achieve their goals.

References

1. Nichol, J., Briggs, J., Dean, J.: PROLOG in education. Educational Review 39(2), 137–146 (1987)
2. Nichol, J., Dean, J., Briggs, J.: Teachers encounter PROLOG. Journal of Computer Assisted Learning 2(2), 74–82 (1986)
3. Hornby, T.: Acorn and the BBC Micro: From Education to Obscurity (2007), http://lowendmac.com/orchard/07/0228.html (February 12, 2010)

4. Coll, J.: The BBC microcomputer: User Guide. British Broadcasting Corporation, London (1982)
5. Williams, C.: Castle bids farewell to RiscPC: Somebody call the fat lady. Drobe launchpad: the archives (2003),
    `http://www.drobe.co.uk/riscos/artifact869.html` (February 14, 2010)
6. Lillingston, J.: Iyonix Ltd. (2008), `http://www.drobe.co.uk/extra/PR07-IYONIXproductiontocease.txt` (February 14, 2010)
7. ARM Ltd. Company Profile: Milestones (2009),
    `http://www.arm.com/about/company-profile/milestones.php` (February 14, 2010)
8. Krazit, T.: ARMed for the living room. CNET News (2006),
    `http://news.cnet.com/ARMed-for-the-living-room/2100-1006_3-6056729.html` (February 14, 2010)
9. ARM Ltd. Corporate Governance: Board of Directors (2010),
    `http://ir.arm.com/phoenix.zhtml?c=197211&p=irol-govboard` (February 14, 2010)
10. Intel Corporate Timeline (2009),
    `http://www.intel.com/museum/corporatetimeline/` (February 14, 2010)
11. Petersen, M.: Review: Premium Softcard IIe. InfoWorld (InfoWorld Media Group) 6(6), 64 (1984)
12. Veit, S.: PC history: Pre-IBM PC Computers (2002),
    `http://www.pc-history.org/` (February 12, 2010)
13. Markoff, J.: Why Apple Sees Next as a Match Made in Heaven, The New York Times, p. D1 (December 23, 1996) (newspaper),
    `http://query.nytimes.com/gst/fullpage.html?res=9F06E1D71331F930A15751C1A960958260` (February 12, 2010)
14. Richards, D.: The new Apple iPad processor the media forgot. Smarthouse: The lifestyle technology guide (February 2, 2010),
    `http://www.smarthouse.com.au/Home_Office/Industry/G6P4K6R7?page=2` (February 15, 2010)
15. Wilson, R.: Apple iPad processor strategy exposed. Computer Weekly (January 28, 2010),
    `http://www.computerweekly.com/Articles/2010/01/28/240104/Apple-iPad-processor-strategy-exposed.htm` (February 14, 2010)
16. Hey, A.J.G.: Supercomputing with transputers—past, present and future. In: Proceedings of the 4th International Conference on Supercomputing (1990),
    `http://delivery.acm.org/10.1145/260000/255192/p479-hey.pdf?key1=255192&key2=0206885621&coll=GUIDE&dl=GUIDE&CFID=75789577&CFTOKEN=52277307`
17. Arabnia, H.R.: The Transputer family of products and their applications in building a high performance computer. In: Belzer, J., Holzman., A.G., Kent, A. (eds.) Encyclopedia of Computer Science and Technology, vol. 39, p. 283 (1998)
18. Charles, D.R., Benneworth, P.S.: Clustering and economic complexity - regional clusters of the ICT sector in the UK. Paper Presented to the OECD Cluster Group Workshop, Utrecht (2000),
    `http://www.oecd.org/dataoecd/34/42/2099353.pdf` (February 14, 2010)
19. Letdorf: A Transputer chip – Wikipedia (2006)
20. Hiranandani, S., Kennedy, K., Tseng, C.: Compiling FORTRAN D for MIMD distributed-memory machines. Communications of the ACM 35, 66–80 (1992)

21. Widas, P.: Introduction to Finite Element Analysis. Virginia Tech Material Science and Engineering (1997),
http://www.sv.vt.edu/classes/MSE2094_NoteBook/97ClassProj/num/widas/history.html (February 13, 2010)
22. Beowulf.org, What makes a cluster a Beowulf (2007),
http://www.beowulf.org/overview/index.html (February 15, 2010)
23. Kim, H., Smith, J.E.: An Instruction Set and Microarchitecture for Instruction Level Distributed Processing, isca In: 29th Annual International Symposium on Computer Architecture (ISCA 2002), p. 0071 (2002)
24. Valiant, L.G.: Bulk-Synchrony: A Bridging Model for Parallel Computation. In: Proceedings of DMCCS, Charleston (1990)
25. Hine, B., Fong, T.W.: Evaluation of the Intel iWarp Parallel Processor for Space Flight Applications. In: AIAA Aerospace Design Conference (February 1993),
http://www.ri.cmu.edu/pub_files/pub4/hine_butler_1993_1/hine butler_1993_1.pdf (February 13, 2010)
26. Smirni, E., Reed, D.A.: Workload characterization of input/output intensive parallel applications. In: Marie, R., Plateau, B., Calzarossa, M.C., Rubino, G.J. (eds.) TOOLS 1997. LNCS, vol. 1245, pp. 169–180. Springer, Heidelberg (1997)
27. Centre for Entrepreneurial Learning (no date) The Cambridge Entrepreneurs: Tudor Brown,
http://www.cfel.jbs.cam.ac.uk/resources/cambridgeents_brown.html (February 12, 2010)

The Birth of Information Systems

Audra Lukaitis, Stas Lukaitis, and Bill Davey

RMIT University, Melbourne, Australia
{Audra.Lukaitis,stas,Bill.Davey}@rmit.edu.au

Abstract. This paper traces the history of the development of information systems degrees in one of the largest Australian universities. A synthesis of documents and transcripts of interviews with the participants shows that information systems grew from an amalgam of existing business courses. The shape of the degrees was initially forged by politics and personality, with a stable and robust curriculum in place after a number of years. This historical narrative shows that university curriculum reflected the significant impact of information technologies in business.

Keywords: information systems curriculum, information systems history, information systems.

1 Introduction

Australia has a long history, relatively, of computing. The CSIR Mk1 (CSIRAC – Council for Scientific and Industrial Research Automatic Computer), built in the late 1940s was Australia's first internally-stored-program computer and is acknowledged to be the world's fourth [1]. People from this burgeoning area were recruited into universities at a very early stage. At the Royal Melbourne Institute of Technology (now RMIT University) Brian O'Donahue from the Commonwealth public service was recruited to head the Data Processing Group. People like Cliff Forrester, who had become engaged by computers after working on EDSAC (Electronic Delay Storage Automatic Calculator) in England as a result of his aeronautical jobs at Farnborough, drifted from some part time teaching at Caulfield Institute (one of the first Colleges to offer formal degrees in information systems) into a full time role at Royal Melbourne Institute of Technology (RMIT).

1.1 RMIT University Beginnings

RMIT is one of the largest universities in Australia. It offers undergraduate and postgraduate degrees in a number of computer-related areas, principally computer science and information systems. The first is quite old in Australian terms. The University opened for business on 7 June, 1887, as the Working Men's College with an initial enrolment of 600 students. By 1988 The College offered classes in technical, business and arts with an emphasis on applied skills relevant to trades including architectural and mechanical drawing, theoretical and applied mechanics, plumbing, carpentry and painting. Studies in arithmetic, algebra, bookkeeping, shorthand, physics, photography and so on.

A. Tatnall (Ed.): HC 2010, IFIP AICT 325, pp. 206–215, 2010.

Currently (2010) there are more than 63,000 enrolled students, including 21,000 international students representing more than 100 countries. The heritage of starting as a union supported college for working men has seen the university orientate itself in such a way where curriculum and applied research are heavily influenced by current industrial practice.

The clear industrial orientation and interest in applied rather than pure research provides an interesting beginning and backdrop to the history of the creation of information systems programs. The story of curriculum development presented in this narrative involves academics who were recruited to the University because of their industrial experience rather than because of their traditional academic experiences such as research, publications and grant success.

1.2 Method

This history was constructed from original documents such as brochures, Faculty of Business handbooks, course guides and internal university documents. Face to face interviews were also conducted with Cliff Forester, Audra Lukaitis and Stasys (Stas) Lukaitis. Cliff Forester was employed by the University after leaving Caulfield Institute in 1965 to help teach in the courses set up by Brian O'Donoghue as part of the then accounting degree. Audra Lukaitis and Stasys Lukaitis worked for different parts of the University, and were brought together by the formation of the new department of business information systems, on the centenary of RMIT's opening, 1987. This means that they were able to observe the evolution of existing courses into degree programs in information systems first hand.

1.3 Setting the Scene – Before the Department

RMIT was considered to be a college of advanced education (CAE) (the same mission as a polytechnic) headed by director Brian W. Smith and associate director David Beanland who subsequently became the vice Chancellor when University status is achieved in 1992 along with the amalgamation with Philip Institute of Technology (PIT).

The history of computing at RMIT commences in 1962 with the lease of an Elliot 803 computer [2]. In 1986 the "Faculty of Business" (subsequently the Business Portfolio and now College of Business) was composed of the Department of accountancy, the Department of administrative studies and the Department of applied economics. In the first instance the Department of accountancy employed some people from industry to deal with the problem of students, particularly postgraduate students, wanting to be able to manage, oversight, implement or participate in the introduction of data processing equipment and systems into their organisations.

Eventually there were enough people to create a data processing group and for a leader of that group to the employed. The first leader was Cliff Forester who became so disenchanted with administrative tasks that he applied for leave to undertake a Masters in computing at the University of Texas. His replacement, Brian O'Donaghue, took the data processing group, now considerably enlarged because of the demand for courses, into the newly created Department of administrative studies.

The group provided significant service teaching throughout the faculty of business as well as offering streams into some degrees. Eventually the Department of administrative studies graduate diploma degrees, starting with first computing program: the Graduate Diploma in commercial data processing. This timeline compares with the creation of the Department of Computer Science in 1981 [2].

1.4 1987 – The Department Is Born

In 1987 the faculty of business saw the demand for information systems courses to be too large for a group within the administrative studies department. The Department of business information systems was created by Tony Adams (who replaced George Sutherland in 1984). Tony Adams had previously headed up a Computer Science group at RMIT, and before that had come from industry where he was the Data Processing Manager for Monash University. By the end of 1984 the Data Processing Group consisted of Tony Adams (Principal Lecturer), Neville Stern and Nigel Thomas (Senior Lecturers), and lecturers Cliff Forrester, Hugh Ballantyne, Philip Crutch and Stas Lukaitis to lead the new department [2].

1.5 1988 – An Undergraduate Degree Is Created

In Australia university places are created by the federal government and partially funded by them. This funding arrangement was especially important for the formation of undergraduate degrees at that time. A new degree could only be created by taking student places away from other courses. RMIT had very successfully run secretarial business studies degrees and graduate diplomas in the 1960s and 1970s and had an outstanding reputation in this area. In the late 1970s the university hierarchy attempted to close the degrees. RMIT had underestimated the "fight back" response of the secretarial studies group who galvanised a large group of alumni, many of whom worked for powerful people in powerful places, including leading business leaders, politicians, and members of Parliament. There was a large "penultimate" city protest/demonstration with news coverage.

The secretarial group consisted of a small number of older (not too far off from retirement) women who had given good service to the Tech (as RMIT was fondly referred to), students and the community. Their educational standards were rigorous and their group was run along "traditional" lines, from another era. There was a strong hierarchy, a very strong commitment to "standards" from which the group did not swerve, there were many students who could never graduate as they could not meet specified required minimum skill standards (which were actually quite "high" by today's standards). In their way they kept up with technology, however were finally stumped by the introduction of the PC. The Tech had to "back down" from the total shutdown of the secretarial group, reviewing their position regarding redundancy.

The secretarial staff mostly had accounting qualifications, two staff members had extensive industry experience but no university qualifications. One particular staff member was also very active in the Victorian Institute of Management, and other business organisations. She also had a large network of important industry contacts that were beneficial to courses, subjects, staff, students and the Tech all round. As a result the old secretarial degree was rebadged as the Bachelor of Business (Office Systems).

The New school, the Department of Computing (later renamed the Department of Business Information Systems) and undergraduate degree and postgraduate diplomas (a one year full-time or two-year part-time for graduates with a three-year university (or equivalent) degree were built on the funding allocated to the secretarial undergraduate four year degree (including the full-time year of work in industry). .

At that time there was no data processing undergraduate degree. So the school was built on the secretarial studies government funding. The funding was split 50/50 between two streams: office systems and business information systems (each stream consisting of 8 full semester units).

1.6 New School Built on Large Compulsory Core Subject Enrolment

Equally important, the school was also built on the large compulsory enrolment of all Faculty of Business students into the computing foundation subject "Computer Applications in Business". At that time over 1,300 students were enrolled in this subject. The whole school taught into this subject, including the Head of School and the Principal Lecturer both on campus and overseas, and every staff member contributed content, direction and teaching, including lectures and laboratory sessions. It was also at this time that the Faculty of Business started to experience a leap in Faculty of Business enrolments. Demand was burgeoning across all business programs.

1.7 Computer Applications in Business Subject – Importance

This subject (course) was always considered very important to the school, as it was the portal to other information systems and business computing subjects offered by the school, also it was an important source of university funding, It was a core subject of all Faculty of Business degrees. Incidentally the content and style of the subject was built on industry consultancies delivered by RMIT's Australian Microcomputers Industry Clearing House (AMIC) – a consulting arm of the University staffed by department academics and other computer specialists.

Although generally the software and other concepts have changed with developments and projected developments in computing in business, the framework is essentially the same as it was 24 years ago. Like the London underground map, the introductory computing subject map put together by the data processing and office systems groups has stood the test of time.

1.8 1988 – The First Bachelor of Business (Business Information Systems) Degree and Growth in Postgraduate Diploma Enrolments

By 1988 the new Bachelor of Business (Business Information Systems) consisted of two major streams: Business Information Systems and Office Systems. The new program extended over four years (of which there was a one year work placement component).

Each stream also offered a one year full-time, or two-year part-time, graduate diploma (a postgraduate qualification) consisting of eight subjects. The jobs the office systems postgraduate course targeted for their graduates were Systems Training Officer, Market Planning Manager, Personnel Officer, Information Manager, Personnel Officer, Projects Coordinator, Change Management Consultant, Office Technology

Consultant, Data Base Systems Officer, International Marketing Specialist and such like. Business Information Systems targeted jobs such as Systems Analyst, Database Programmer/Analyst/Administrator and information technology consultant and advisor.

The Postgraduate Diploma in Secretarial Studies continued (with large numbers of students still applying). The Postgraduate Diploma in Commercial Data Processing also still continued (with large numbers of students applying and growing). 1988 also saw the launch of the first Bachelor of Business (Marketing) in the Faculty of Business. In addition, application had been made to the appropriate authorities to change the names of several courses currently offered by the faculty. It is pertinent to note that at that time all programs were fully funded by the Government and places were offered to students on a competitive basis, students did not pay fees.

1.9 Consulting and Industry Engagement

The department and its academic staff had arisen from demands of industry for job specific training. This was (and is) reflected in working conditions and priorities. Until recently PhD qualifications were seen as irrelevant as the role of the college of advanced education was "*more applied and less research-oriented than universities*" [3:121]. The Martin Report [4] tabled in 1964-65 recommended the creation of colleges of advanced education rather than the expansion of universities to cope with the demand for more universities places. The mission of the College of Advanced Education (CAE) was therefore very different from that of a university.

Rather than research, RMIT, CAE academics were required to teach or consult into industry and maintain close links with industry. In her interview Audra Lukaitis describes this priority

> "*It was a very heavy load to work in industry as a consultant and then work full-time as an academic. Nevertheless this experience was to prove absolutely invaluable to me and has informed my whole approach to computing education. To industry I was able to bring fresh ideas, concepts and theories garnered from research in my fields, and from industry I was able to bring back real world case studies that informed all assignments, approaches and curriculum content. I was able to see firsthand what really went on in industry as a participant and bring it back to the classroom.*"

In an interview, Stas Lukaitis remarks

> *Perhaps it might be of interest to note that AMIC – Australian Microcomputer Industry Clearinghouse - was created by an enterprising Tony Adams and others in response to the "sudden" arrival of microcomputers onto the scene. The demand for short courses and consultancies on the use and deployment of micros, their software e.g. – Visicalc, Lotus, MultiMate, WordStar, Dbase was huge and AMIC blossomed. Several BIS staff were recruited from the ranks of the consultants who worked at AMIC (e.g. Peter Viola who was a Manager at AMIC for many years).*

> *Academics who worked at AMIC were actually paid at standard industry consulting rates. It was this that created a great deal of angst from the Chancellery who demanded that academics stop being paid so much. That was the start of the demise of AMIC, the birth of the MDC and the end of academic input into University education engagement.*

1.10 International Demand

Right from the formation of the Department of Business Information Systems (now School of Business Information Technology and Logistics) the founding Head, Tony Adams along with others in the Faculty of Business began to pioneer and establish international education initiatives onshore and offshore, namely in Malaysia, Singapore and Hong Kong, particularly Barry Cooper who setup the original Malaysian operation at Taylors College.

Staff were now required to teach overseas (in addition to consulting in industry), which certainly added another challenge, but without a doubt yet another valuable dimension. Groups of students from around the world appeared in classrooms, this brought many new challenges for which many staff were unprepared - academic standards, assumptions, English language proficiency, writing and comprehension, extremes in diversity, cultural differences - all had to be dealt with. The international student market was grown very aggressively and successfully by Barry Cooper from Accountancy, Colin Bent and Tony Adams (onshore and offshore). They were initially the ones who had the vision to expand RMIT's activities globally, they were the pathfinders. Once again, teaching overseas was considered to be "above load", so academics were paid separately for overseas teaching while still putting in full time back home, nevertheless all academics were required to teach overseas.

2 Life and Death of Office Systems

New Office Systems subjects: Principal Lecturers were responsible for the main groupings of staff who delivered subject (courses, subjects = programs, courses). Department of Administrative Studies: Tony Adams, James Hearne, James Hurley.

Table 1. 1986 Undergraduate Programs

1986 - Undergraduate Programs				Special Comment
Course Name Bachelor of Business in the fields of:	**Department**	**Program Length**	One Year Industry Experience Supervised professional practice	
Accountancy	Accountancy	4	√	
Business Administration	Administrative Studies	4	√	
Local Government	Administrative Studies	4	√	
Property	Applied Economics	4	√	
Public Administration	Administrative Studies	4	√	
Transport	Applied Economics	4	√	
Secretarial Studies*	Administrative Studies	4	√	No new students may enrol as this course is undergoing review

* Students who were still currently undertaking this course would complete their studies in accordance with the 1984 Advanced College Handbook.

Table 2. 1986 Postgraduate Programs

1986 - Postgraduate Programs	
Graduate Diploma (Prerequisite – 3-year undergraduate degree)	**Department**
Commercial Data Processing (Grad. Dip. Comm. Data Proc.)	Administrative Studies
Finance (Grad. Dip. Fin.)	Accountancy
Internal Auditing (Grad. Dip. Int. Aud.)	Accountancy
Management (Grad. Dip. Mgmt)	Administrative Studies
Organizational Development (Grad. Dip. Org. Dev.)	Administrative Studies
Real Estate Development and Investment (Grad. Dip. R.E. & Inv.)	Applied Economics
Secretarial Studies (Grad. Dip. Sec. Stud.)	Administrative Studies
Taxation (Grad. Dip. Tax.)	Accountancy
Accounting and Auditing	Accountancy
Health Services	Administrative Studies

Table 3. 1987 Postgraduate Programs

1987 - Undergraduate Programs				
Course Name Bachelor of Business in the fields of:	**Department**	**Program Length**	**One Year Industry Experience Supervised professional practice**	**Special Comment**
Accountancy	Accountancy	4	√	
Business Administration	Administrative Studies	4	√	
Local Government	Administrative Studies	4	√	
Property	Applied Economics	4	√	
Transport	Applied Economics	4	√	

Dean Noel Anthony appointed Foundation Dean of the Faculty of Business in 1977, commencing as a Lecturer in Accountancy in 1963 after many years service with the State Bank of Victoria. Appointed Head School of Business Studies in 1976, was also Acting Associated Director, Advanced College. Noel Anthony was responsible for the new extensive computer facilities within the Faculty; he was actually the one who encouraged the Faculty of Business Computing Group to acquire and deploy the new microcomputers and funded the setup of the first PC labs in Business. Noel Anthony Retired in 1986, with the new Dean Dr. J. Milton-Smith appointed 1986, commencing in 1987.

Department of Administration staff (relevant to future new school): Tony Adams, Hugh Ballantyne, Helen Carthew, Philip Crutch, Cliff Forrester, Eileen Gueho, James Hearne, Stas Lukaitis, Fiona Petersen, Margaret Petherick, Neville Stern, Nigel Thomas, Audra Lukaitis.

3 1987 – Centenary Year

By 1987 the undergraduate degrees in Public Administration and Secretarial Studies were gone.

In postgraduate programs 2 were discontinued and 3 were renewed, rebated, or new.

1992 was the declaration of the Act establishing RMIT as a university and also the amalgamation of Philip Institute where the department acquired a new head Ken Millar and a number of new staff.

1984-1990   Tony Adams Head
1990-1993   Neville Stern Acting Head)
1993-2001   Ken Millar (Head)
2002-2002   Tom Yardley (Acting Head)
2002-2003   Kevin Adams (Acting Head)
2003-2005   Caroline Dowling (Head)
2005-2005   Barry McIntyre (Acting Head)
2006 -   Brian Corbitt (Head)

Remaining PIT people: Bill Davey, Elspeth MacKay, Lisa Keay

Table 4. 1986 Postgraduate Programs

1987 - Postgraduate Programs		
Course Name	**Department**	**Comments**
Graduate Diploma (Prerequisite – 3-year undergraduate degree)		
Transport and Distribution (Grad Dip Transport & Distribution)	Applied Economics	New
Commercial Data Processing (Grad. Dip. Comm. Data Proc.)	Administrative Studies	
Finance (Grad. Dip. Fin.)	Accountancy	
Internal Auditing (Grad. Dip. Int. Aud.)	Accountancy	Discontinued
Management (Grad. Dip. Mgmt)	Administrative Studies	
Organisational Development (Grad. Dip. Org. Dev)	Administrative Studies	
Real Estate Development and Investment (Grad. Dip. R. E. & Inv)	Applied Economics	Discontinued
Property (Grad. Dip. Prop.)	Applied Economics	New
Secretarial Studies (Grad. Dip. Sec. Stud.)	Administrative Studies	
Taxation (Grad. Dip. Tax.)	Accountancy	
Accounting and Auditing	Accountancy	
Health Services Management (Grad. Dip. HSM)	Administrative Studies	New

4 Conclusion

This historical record in this case of RMIT shows a number of very interesting outcomes.

The demise of secretarial studies and the degree programs that emerged with the creation of the information systems department seem illogical. Managers continue to have executive assistants and support staff who need to have particular skills in using technology and an understanding of information systems. The argument used against programs for these types of people seems to have been based upon a belief that secretarial studies consisted of training in technology that had been completely replaced by the computer. This argument concentrates on the replacement of the typewriter and shorthand with the word processor and dictaphones.

Another outcome may be reflected in the nature of information systems curriculum. At RMIT computer-based courses took two paths right from the beginning. Computer science was developed from the machine, its design, and programs that could run on it. Computer science was based in the disciplines of mathematics and applied physics and, as such, could be seen as valid integration of those two areas. The Computer Science department arose from the ashes of the Department of Mathematics and Computer Science, originally just Mathematics. It's of interest that most Computer Science people originated from maths and science where it was based. Much of the curriculum was scientific and mathematical with lip service (and contempt) paid to business and "commercial data processing" which was seen as particularly unsexy. CS degrees in the 1970's had compiler design, the solutions of differential equations, operations research, calculus, artificial intelligence and a peculiar form of software engineering. Your comments about IS arising from business demand is such a compelling factor that it has become lost.

On the other hand information systems arose from the demand from organizations to provide education and training for staff required to implement new technologies. In the late 1960s and early 1970s there was a plethora of courses aimed at producing workers in business capable of playing a part in the data processing of the business. The early programs offered by the data processing group and in the information systems department were heavily skewed towards postgraduate education. Vast numbers of students enrolled in postgraduate programs part-time so as to gain the computing skills and education needed in their new jobs.

It is no coincidence that the accounting department was an early sponsor of these courses and the creation of the data processing group. The Accounting Department would be the first to understand the changes happening in business due to introduction of computers in data processing.

When the first head of the information systems department came from computer science he may have been tempted to introduce some computer science rigor into the courses. His employment practices clearly indicated that he understood the difference between computer science and information systems. He hired senior people from IBM, HP and other businesses who would understand the needs of the graduates of these courses. This meant that the courses developed by the information systems department were heavily biased towards current practice in industry. The programming languages Cobol rather than FORTRAN. Systems subjects were prominent in many subject names used the words data processing.

This paper recounts the history of the single university's information systems degrees. It would be interesting to look at the history of other information systems programs to see if similar backgrounds have produced an environment similar to that of the current information systems programs at RMIT.

References

1. Tatnall, A., Davey, B.: Streams in the History of Computer Education in Australia. In: Impagliazzo, J., Lee, J.A.N. (eds.) History of Computing in Education, pp. 83–90. Kluwer Academic Publishers/IFIP, Assinippi Park (2004)
2. Tatnall, A.: Curriculum Cycles in the History of Information Systems in Australia. Heidelberg Press, Melbourne (2006)
3. Karmel, P.: The Role of Central Government in Higher Education. Higher Education Quarterly 42(2), 119–133 (Spring 1988)
4. Martin, Sir L. (Committee Chair): Report of the Committee on the Future of Tertiary Education in Australia the Martin Report, 3 vols. Australian Universities Commission, AGPS, Canberra (1964-1965)

The Monash University Museum of Computing History: Ten Years On

A. Barbara Ainsworth, Chris Avram, and Judithe Sheard

Faculty of Information Technology
Monash University
{Barbara.Ainsworth,Chris.Avram,Judy.Sheard}
@infotech.monash.edu.au

Abstract. The Monash Museum of Computing History was established at Monash University, Melbourne, Australia in 2001. This university museum has a growing collection of early computer equipment and a permanent exhibition tracing the development of computing technology within an Australian context and particularly related to computing at Monash University. The Museum has been evolving since its inception with greater definition of its collection policy, defined collection management and its role as a repository for computing history and the dissemination of its research. This paper gives an overview of the origins of the Museum, current activities and future directions.

Keywords: history of computing, computing museum.

1 Introduction

The Monash Museum of Computing History (MMoCH) is located at the Caulfield campus of Monash University, Melbourne, Australia. The Museum is an initiative of the Faculty of Information Technology and combines a physical exhibition, a collection of original material and research activities. It was started in 2001 with a brief to collect original computing material around the University and develop an exhibition. The role of the Museum has expanded with continued collecting and research but we are currently exploring new methods of presenting this information to both students and other interested users through public programs. It is appropriate to provide a review of the first ten years surveying the origins of the Museum, its collection and research activities, exhibition program and public programs as well as discussing the future directions now in planning. This paper gives an overview of the origins of the Museum, current activities and future directions.

1.1 Background of Museum

The concept of a museum at Monash University was inspired by a speech by Max Burnet to staff from the Faculty of Information Technology in 2001. Max Burnet is well known in Australian computing history research as a major collector of early computer material and he is an authority on the development of computing in Australia. His speech highlighted the rapid change in computer technology and the lack of

A. Tatnall (Ed.): HC 2010, IFIP AICT 325, pp. 216–227, 2010.

preservation of early material as most institutions replace old technology on a regular basis. Older equipment was generally seen as a nuisance, something occupying valuable space and disposable. Dr Judithe Sheard, Senior Lecturer in the Caulfield School of Information Technology, heard Max's speech and felt the significance of older equipment used at Monash University should not be lost. She approached the Faculty and gained permission to develop a museum that would preserve this early technology and produce a physical display.

There have been computing activities at Monash University and other institutions that have amalgamated with the University since the early 1960s[1]. The University created the Faculty of Computing and Information Technology, now called the Faculty of Information Technology, in 1990[2]. Remnants of past installations and equipment were spread across different departments and locations. The amalgamation of the separate teaching institutions led to the existence of a number of historical computing installations and equipment being used at the different schools and departments. Gradually a large variety of objects, documents and photographs were assembled and a display was established in October 2001 at the Caulfield campus. This was the first stage of the Museum and the Museum has since continued to collect material and improve its physical display as well as develop its museological practices.

Dr Sheard continues her involvement as Director and Sarah Rood, who was working for the Faculty, was seconded to work on the Museum relocation during 2004 and 2005. Together they produced the current major permanent exhibition. An advisory committee was established to give both computing and museum advice. In late 2006 Barbara Ainsworth took up the position of part-time Curator and started formally cataloguing the collection with supportive research. Chris Avram has been an Honorary Consultant since the beginning of the project. Several other academic staff continue to provide research and practical assistance when required. Max Burnet has maintained his connection with the Museum, providing both advice and loan material. Museum personnel have continued to develop both the Museum's practices and exhibitions as well as produce supportive activities and publications.

The Museum has received ongoing funding for collection management and exhibition development. Professor Ron Weber, Dean of the Faculty, has been generous in his continuing interest in this project.

1.2 Purpose of the Museum

In its original inception, the Museum was established to collect and preserve superseded computing material around Monash University and produce a small didactic exhibition. Museum staff then developed a number of policies to clarify the role of the Museum beyond its initial aims and give an explicit set of objectives as well as a Collection Policy. This set of objectives is necessary to define the Museum's purpose and help shape the direction of future developments and ongoing museum practices. It

---

[1] Monash University started at the Clayton campus with later amalgamations adding several different campus locations. For a brief history see the Interactive pictorial timeline of Monash University [1].

[2] The Faculty of Information Technology represents the current culmination of the gradual development of computing studies at Monash University. Rood gives a summary of these events in her Introduction [2].

is a small museum with limited resources and this makes it necessary to have a clear set of objectives. The role of the Museum has been defined with a mission statement to collect and preserve original computing items associated with Monash University and research their provenance, to undertake research on the development of the Faculty and collect biographical material relating to people involved with the Faculty. The Museum also collects supportive material for reference on the history of computing in Australia. It is a primary function of the Museum to make the collection and research material accessible for educational and research purposes. This material is made available to the academic community, students and the general public through exhibitions and lectures. Researchers can also access original material not on display. The museum provides access to research material through publications and the website. The Museum policies are available on the Museum website [3].

The definition of the Museum's role provides a framework for understanding the different areas of activity in the Museum and our current view of the future direction of the Museum. The Museum's primary function is to collect original material and information and present this material in an accessible form to the academic and general community. The different areas of the Museum's work are seen as interactive, using original material and research to provide resources and material to be presented in virtual or physical environments. Currently information is presented through exhibition text, handouts or educational sheets, as well as published material and the website research. The Museum has been developing different aspects of this model.

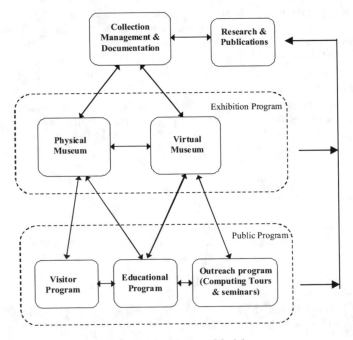

Fig. 1. The Museum Model

2 Collection

A primary aim of the Museum is to acquire original artifacts associated with the development of Monash University's use of computer technology for both administrative and teaching purposes. The collection has a wide variety of calculating and computing material that reflects the changing technology used by staff and students over the last forty years.

The University was given its charter to start a new institution in 1958 and the acquisition of computing equipment was given an early priority. After long consideration by a special committee, the University ordered a Ferranti Sirius computer from the English company Ferranti Ltd. It was shipped from England and installed by November 1962. Further investigations revealed that were actually four different Sirius computers at the current campuses of Monash at different times over the 1960s. These four computers were located at different sites in 1962 with two Sirius computers at the Melbourne Computer Centre operated by Ferranti Ltd, one at the Clayton campus of Monash University, and one at a commercial research operation run by ICIANZ, now called ORICA, at Ascot Vale. Through loan, donation and sales all four were located at some point at Monash University. One of these machines managed to survive in storage at the Clayton campus and was transferred to the Museum at the Caulfield campus in 2005. It is now a major exhibit in the permanent display. The MMoCH Ferranti Sirius is displayed in a dedicated showcase and includes the 1000 word CPU and an additional 3000 word memory cabinet along with a suite of I/O (input/output) equipment. Further information on the history of the Ferranti Sirius at Monash University is available from an article by Barbara Ainsworth [4] accessed from the Museum website [3].

The Ferranti Sirius was one of the larger items found around the campuses of Monash for the Collection but other interesting items appeared out of storage. An early calculating machine called The Millionaire had been housed in the Maths Department and was donated to the Museum. The Millionaire was based on a design created by the Frenchman Leon Bollee and patented in Germany in the early 1890s by Otto Steiger. The design was further improved by Hans W.Egli in Switzerland, and was manufactured in Zurich from 1893. All machines were given an individual identifying number. They are built mainly of cast iron and brass and weigh about 33 kilograms. Although once commonly used for mechanical multiplication in business, there are now about only 22 examples remaining in Australia. The machine in the MMoCH collection is numbered 913 and was probably produced during the period 1898-1900. For further reading on the Millionaire see John Wolff's Web Museum [5].

Others items in the collection reflect the teaching role of the university. This includes an example of the computer system which was used in the MONECS project. In the late 1970s a school could only afford one computer, so how could 30 children run a program with only one card punch machine and a one hour lesson period? The answer was provided by an ingenious system developed by Monash University. The class would write their programs on mark sense cards, and each deck could be run on the MONECS computer in a few seconds. That way, every student had a chance to run a program in a one hour lesson. The MONECS (Monash Educational Computer System) scheme was a project administered by the Monash Computing Centre and developed by Professor Cliff Bellamy and his associates as a teaching tool for

Monash students as well as off-campus students to develop computer programming skills. The system ran on a PDP-11 which could be a stand-alone system accepting input from cards. This computer was mobile and could be sent out to schools and TA-FEs and other tertiary institutions. There were over 30 outside users in the early 1980s. Monash was running 12 PDP-11 and Spectrum systems in 1984[3] .Students without access to a computer on-site could use mark-sensed cards which were marked by the students at their institution and the cards were then sent to Monash Computer Centre to be run at quiet times, and then returned to the outside school. The MONECS is part of an autobiographical display in the Museum on the work of Professor Cliff Bellamy.

Recently the Museum was able to co-ordinate the return of a computer used in the MONADS project [6]. The MONADS Project was initiated by Professor Leslie Keedy in 1976 at Monash University; it was an umbrella project for related research activities concerned with design of computer systems. This computer design and development project was undertaken in the Department of Computer Science, Clayton campus into the mid-1980s by Professors Leslie Keedy, John Rosenberg and David Abramson. Students and staff continued this research at Monash University until 1985 and then at other institutions. The MONADS I and II computers consisted of a series of circuit boards which were installed into a Hewlett-Packard HP-2100A minicomputer. Professor Chris Wallace and Robert Hagan had already modified the HP-2100A adding a virtual memory system. The MONADS projects used this virtual memory system as a base for further developments. Professor Keedy went to Darmstadt in Germany and continued the project. One of the "MONADS PCs" has recently been returned from the University of Ulm in Germany to the Museum. The machine was designed by Professor David Abramson and Professor John Rosenberg in the mid 1980's while they were academics in the Department of Computer Science at Monash University.

The Museum collection has been catalogued and now numbers over 700 items. New material is regularly offered to the Museum. Currently we are creating a database with these records. John Sheard, an Access database developer, has volunteered his expertise and created an Access database specifically for this purpose. The material has also been classified with an in-house classification system. A large number of items have been placed on permanent exhibition with the remainder in storage. Material in storage is accessible upon request to the Director and is used for lecturing to visiting groups. The Museum does occasionally lend material for teaching purposes.

3   Stages of Exhibition Work

3.1   First Exhibition

The Museum opened its first exhibition in October 2001 with a display of computing material set in a chronological format. The display was entitled 'A Digital Evolution',

---

[3] References on this project are held in the Monash University Archives. Papers include internal memo by Cliff Bellamy 9 May 1984, also printout of users in1984 on file in MU Archives MON 935 1999/6 MONECS 1981-82. See article Monecs Monash Educational Computer System by Cliff Bellamy 13 July 1984 in MU Archives MON 935 1999/6 MONECS 1981-82.

and was an exhibition of computing history. The display was located at the Caulfield campus. This display included a mixture of computing equipment used on campus and some loan items ranging from slide rules and calculators to card readers and early Apple personal computers. Max Burnet placed his example of the computer PDP 9 on display. It was originally installed at La Trobe University, Melbourne in 1967. A team of interested staff, students and outside advisors contributed to the initial development under the direction of Dr Sheard. Professor David Abramson, then Head of School, was very supportive of the project and organized departmental funding. Max Burnet provided both his knowledge and loan items for the first display. He developed the initial display and wrote the text for the display signage. The display labels were then designed by an art and design student. The technical people in the School helped move the equipment.

A temporary display, entitled "Ways and Means of Computing", was also held at the Monash Science Centre, Clayton campus, Monash University in 2002. The display was curated by Dr Sheard with the exhibition content and research developed by Dr Selby Markham. While the display featured original computing material, it also included a number of interactive exhibits to demonstrate concepts behind computing, including a Mechanical Binary Counting Device and model slide rules.

3.2 Permanent Exhibition, Caulfield Campus, Monash University

The current permanent physical display is in the foyer of Building B, Caulfield campus and was officially opened on 11 May 2005. The University received generous funding and support from a number of sources including Multimedia Victoria, Museum Victoria, Australian Computer Society, IKEA, The Age, Floorbotics, Microsoft and Holden Ltd.

The exhibition has four main areas of display. The introductory area features the concept of calculating by external means. The early calculating equipment highlights the problems of mechanical computation. There have been different solutions to this problem and the display features a number of mechanical counting devices, electro-mechanical calculators and the hand-held electronic calculators of the 1970s. There is also a variety of slide rules ranging from the common academic slide rules to specialized devices. Once an essential tool for all mathematicians and engineers, these slide rules are completely unknown to our student visitors. These forms of calculating are a precursor to the use of stored memory computers.

The major display area is devoted to the development of computers with examples used since the late 1950s through to the modern era of ubiquitous and mobile computing. The range of computers shows the commercial sized operations, including the PDP-9 (on loan from Max Burnet) and a console from the computer model IBM 360/30, to a range of early home computers from BBC, Commodore, Acorn and Atari. The Apple Macintosh range is also represented with an APPLE II and an APPLE LISA on display. The computers are displayed in a chronological sequence but a major feature is a pictorial background to the computers. The large range of images represents contemporaneous events to the technological developments. This places the computers within a sociological framework of current events and influences in Australia.

A dedicated showcase features the Ferranti Sirius which was installed at Monash University in 1962. The basic Sirius CPU had 1000 words of store but could be expanded with extra memory units with a 3000 word store. The Sirius on display has the CPU with one extra memory cabinet. The Sirius uses acoustic delay line memory. All input was through punch tape. The showcase also includes a Ferranti paper tape reader, Ferranti Westrex paper tape punch, a Creed teletype and a Creed paper tape reader set on a table. The display is supported by a short b/w film which was produced for the British Railways by Mr.S.E. Fargher in 1963. Fargher's film explains the process of using the Sirius to solve a problem in simple language. The MMoCH transferred the 4000 word machine to the Caulfield campus in 2005. A special committee was formed to help the museum prepare the machine for display and it was relocated to Building B foyer. The committee members were volunteers including Peter Thorne, Judy Hughes, Jurij Semkin and John Spencer. They cleaned the various parts of the Sirius. These volunteers had experience with restoring CSIRAC for the Melbourne Museum. The real-time clock was restored to working order by Chris Avram and Bruce Gilligan.

The fourth part of the display features biographies of people associated with computing at Monash University through management, teaching or research. The first biographies examined the work of Professor Chris Wallace, Professor Cliff Bellamy and Dr Andrew Prentice. Professors Wallace and Bellamy made significant contributions to the development of computing at Monash University. Cliff Bellamy had a long association with the Monash Computer Centre in his role as Director from 1964 and then became the first dean of the current Faculty of Information Technology. Professor Chris Wallace was Foundation Professor of the Department of Information Science which was established at Clayton campus in the Faculty of Science in 1968. More recently in April 2008 the display was updated to include material on the work of Professor John Rosenberg and Professor David Abramson. These biographies feature equipment and publications as well as photographic and text displays.

3.3   Multimedia Project with Arts Faculty

Teaching the history of computing is widely accepted as an important way of helping IT students to better understand their field. This was acknowledged in 1991 with the inclusion of history modules in the ACM/IEEE curriculum for computer science [7]. While the Monash Museum of Computing History is a valuable resource for this purpose, MMoCH staff are collaborating with staff from the Faculty of Arts on a Virtual Museum Project (VMP) to further develop the MMoCH resource by bringing to life the physical exhibits with computer-generated animation. Geoff Berry from the History Department developed a concept to connect different technology over a timeframe. Daniel Simmonds and Dr Tom Chandler from the Multimedia Department, Berwick campus, provided the technical multimedia component of the project.

The first stage of this project has been completed with the production of a short video which covers a period from the mid 60s to the mid 70s. The development of computer technology is illustrated through a story of space exploration. The 'virtual tour' takes in the Ferranti Sirius, the PDP-9, and the hand-held HP-65 calculator. The animated narrative places the technology in the context of current events, helping students understand the influences of computers on society and the factors that influenced the development of computer technology. The video can be explored from a number of different aspects relevant to a broad range of computer science courses. Further description of this resource is available elsewhere [8].

4 Publications and Research Activities

4.1 History Publications

The Faculty of Information Technology, Monash University commissioned a history of the Faculty after the idea was discussed at the launch of the 2001 MMoCH display. Professor John Rosenberg, Dean of the Faculty from 1997-2003, agreed that the period of time leading up to the formation of the Faculty coincided with significant changes in the approach to teaching information technology and the Faculty should support a history publication. A steering committee was formed to guide the project. Sarah Rood, as well as being Curator for the MMoCH, was given the contract to write the history. The book follows the development of computing at Monash University, starting with the Monash Computer Centre and the beginnings of the teaching school in the late 1960s to the establishment of a separate faculty in 1990. The author interviewed a large number of people, including former students and staff, involved with computing at Monash over the past thirty years. The book From Ferranti to Faculty was published in 2008 [9].

In 2008 the Museum was represented in inroads, SIGCSE Bulletin with a two part article about the Museum. This was a great opportunity to raise the profile of the Museum and highlight some of the interesting parts of the collection to a wide audience [10, 11].

4.2 Research

Internal research has been focused on material within the collection. This has been particularly important for establishing the provenance of the Ferranti Sirius computer in the collection. It came from the Clayton campus into the collection in 2001, but we were aware that Monash University had donated another Sirius to Melbourne Museum in 1975. Director Judithe Sheard had also received a number of communications from older staff members with their recollections about the Ferranti Sirius at Monash University. These recollections proved to be quite confusing and it became apparent that they referred to several different installations. After examining historical resources, including Monash University Archives and ORICA Library and Archives, we now understand that there were four different Sirius computers in Melbourne during the early 1960s and all four were located at Monash University at different points in time. We have also been fortunate to have received assistance with technical and historical information from members of the British Computer Conservation Society. This collaboration has led to the inclusion of an article on the Sirius at Monash University in the magazine RESURRECTION, The Bulletin of the Computer Conservation Society [12].

5 Public Program

The MMoCH has developed a number of public activities to raise the profile of the Museum and introduce aspects of the collection and display to a larger audience.

5.1 Museum Website

The Museum has developed a website to provide information on the Museum and its operational matters. The site also allows access to some of the current research projects being undertaken at the Museum. A number of leaflets on specific topics are also posted. These include a brief history of computing at Monash University, the Ferranti Sirius computer in the collection and notes on early calculators. Another aim of the site is to provide resources for further study and indicates other locations for information on computing history. The website can be found at http://www.infotech.monash.edu.au/about/projects/museum/

5.2 Seminar with Gordon Bell – Bits and Bytes

In May 2008 the MMoCH hosted a seminar with guest Gordon Bell giving an address on his view of current uses for the recording of personal information as well as a seminar on the Computer History Museum, California. Gordon Bell spoke about his latest work in developing the MyLifeBits project which is an experiment in using multimedia to record every aspect of his daily life. The MyLifeBits project seeks to record these in a personal transaction processing database [13].

Gordon Bell has a long career in the computing industry starting with 23 years (1960-1983) at Digital Equipment Corporation. He has been involved with a number of advisory committees, including the Sector Advisor Committee of the ICT Division, CSIRO, encouraging the development of new computer technology. Currently Gordon is a principal researcher in the Microsoft Research Silicon Valley Research Group. In 1979 Gordon Bell and Gwen Bell were co-founders of the Computer Museum, Boston, Massachusetts. This museum developed in Boston but the collection was later transferred to California and became the Computer History Museum in Mountain View, California. Gordon's background gives him a unique perspective on computer technology and the role of computer museums.

This seminar was a joint project between the Monash University's MMoCH, eResearch and the Centre for Community and Organisational Informatics (COSI). The seminar drew an interested group from the academic, museum and professional IT areas. The success of this event has encouraged the MMoCH to organize more events in the future.

5.3 Computer History Tours

In 2008 Associate Professor Graham Farr approached the MMoCH to collaborate on a new initiative to create a history of computing tour using the Melbourne tramway system to connect different museum exhibitions and sites with a general theme of computer history. The tour groups usually number between 30-40 people with Graham Farr, Judithe Sheard, Chris Avram and Barbara Ainsworth providing a narrative at different sites. The tour can also be undertaken as a self-guided individual activity using tour notes from the website. A list of suggested sites and more extensive notes can be accessed through the Computing in Melbourne: A Historical Tour website [14].

These tours commence at the MMoCH exhibition at Caulfield campus and then the tour usually follows the tram route from Caulfield, into St Kilda Road, then the Royal Botanical Gardens, into the City precinct. The tour then continues with a visit to the computer exhibition at Melbourne Museum featuring CSIRAC, the only remaining 1st generation stored program computer [15, 16]. The tour then concludes with a visit to Melbourne University to discuss the first Internet connections in Australia. The different sites can be varied for the tour. The sites do not necessarily have original computing equipment but are used as an introduction to a related computing history topic. Usually these tours, which are free, are run on a weekend but are also available by special arrangement. This has been an interesting exercise for both the guides and participants. Different people attend the tours including current academic staff, postgraduate and undergraduate students as well as many retired IT professionals. The conversations stimulated by the tour notes and information have introduced many new topics and sources of information for further research.

6 Future Directions

The Monash Museum of Computing History future development is directed towards creating a virtual museum presentation. This on-line educational environment will be a learning space where students and other interested users can explore the history of computing with a special focus on Australian computing history and Monash University's part in this history. This facility will provide another component in the Museum's different physical and virtual presentations and link its research program and collection with on-line visitors who can participate in different forms of access for varying purposes. It will be a virtual space on the Internet that provides a multimedia portal into the history of the development of computers including both the technological advances and the social implications of this technological change within an Australian context. The project will use a range of resources to give the visitor a variety of information sources including in-depth text-based research information or interactive participation using different media including three-D imaging and original moving or still images. The portal will also give access to view the physical collections and displays at Monash University. The site will provide different levels of access that address the specific needs and internet literacy of the visitor. It is anticipated that the project will be used as a resource for different levels of academic users, remote visitors and general interest users and respond to their differing needs. Students would also have educational resources and teaching material available to support their studies. Another feature of the Museum will be an interactive component where visitors can contribute their own knowledge and experiences to our reference base.

The basic level of access would be through a timeline which gives linkage points to topics mentioned in the timeline text. The timeline provides all visitors with a summary of the narrative of the overall topic, the beginnings of stored memory computers within the context of the history of calculating and the use of numbers with special reference to Australian developments. The timeline is then the starting point for a visitor to explore the main points further. By clicking on a topic, the visitor is then offered a range of sub-topics presented in different formats. They can read text or activate presentations of original footage or still images. The preliminary multimedia project (described earlier) has been investigating the development of imaging early technology and activating the images in a realtime situation to recreate the technology in a contemporaneous context. For specialized visitors, an access point is provided for collection searches on the MMoCH physical collection. This would show the provenance and physical attributes of these objects as well as images. A visitor can also have a virtual tour of the current physical display at the museum.

This project is still in the planning stages but staff have had preliminary meetings with academic advisers from the History Department and the IT Faculty, Monash University. The Museum hopes to provide a portal on the history of computing which exploits the skills of different faculties to interpret the Museum's physical collection and research. The Internet will allow the Museum to reach a wider audience and provide a resource for different levels of students as well as the general user.

7 Conclusion

The Monash Museum of Computing History is a relatively new university museum but, through the use of modern media, it aims to provide information and resources to a wide community of users. The physical collection gives staff and students a base for research in a number of areas including technological developments and historical interpretation of the impact of these developments. Future plans will expand the outreach of the Museum from the immediate university environs to a much larger audience through the Internet.

References

1. Our History, achievements and milestones: 1990 - Caulfield and Peninsula campuses established (1990),
   http://www.monash.edu.au/timeline/1990-caulfield-peninsula.html (accessed: February 25, 2010)
2. Rood, S.: From Ferranti to Faculty: Information Technology at Monash University, pp. ix-x. 1960 to 1990 Monash University ePress, Melbourne (2008)
3. Monash Museum of Computing History,
   http://www.infotech.monash.edu.au/about/projects/museum (accessed: February 25, 2010)
4. Ainsworth, A.B.: The Ferranti Sirius computer at Monash University (2008),
   http://www.infotech.monash.edu.au/about/projects/museum/papers/first-computer-at-monash-university-v4.pdf (accessed: February 25, 2010)
5. John Wolff's Web Museum,
   http://home.vicnet.net.au/~wolff/calculators/Egli/Egli.htm (accessed: February 25, 2010)
6. The Monads Project, http://monads-security.org/ (accessed: February 25, 2010)

7. Lee, J.A.N.: History in the computer science curriculum. ACM SIGCSE Bulletin 28(2), 15–20 (1996)
8. Berry, G., Sheard, J., Quartly, M.: A Virtual Museum of Computing History: an educational resource bringing the relationship between people and computers to life. Paper Submitted to Australasian Computing Education Conference (2011)
9. Rood, S.: From Ferranti to Faculty: Information Technology at Monash University. 1960 to 1990 Monash University ePress, Melbourne (2008)
10. Ainsworth, A.B., Sheard, J., Avram, C.: Monash Museum of Computing History Part I. SIGSCE Bulletin Inroads 40(2), 31–34 (2008)
11. Ainsworth, A.B., Sheard, J., Avram, C.: Monash Museum of Computing History Part II. SIGSCE Bulletin Inroads 40(4), 31–34 (2008)
12. Ainsworth, B.: The Ferranti Sirius at Monash University, RESURRECTION. The Bulletin of the Computer Conservation Society (44) (August 2008), http://www.cs.manchester.ac.uk/CCS/res/res44.htm (accessed: February 25, 2010)
13. Gemmell, J., Bell, G., Lueder, R.: MyLifeBits: A personal Database for Everything. Communications of the ACM 49(1), 44–50 (2006)
14. Computing in Melbourne: A Historical Tour, http://www.csse.monash.edu.au/~gfarr/tour/
15. CSIRAC: Australia's first computer, Museum of Victoria, http://www.museum.vic.gov.au/csirac/ (accessed: February 25, 2010)
16. McCann, D., Thorne, P.: The Last of the First. CSIRAC: Australia's First Computer. University of Melbourne (2000)

Author Index